RHETORIC IN THE FLESH

Rhetoric in the Flesh is the first book-length ethnographic study of the gross anatomy lab to explain how rhetorical discourses, multimodal displays, and embodied practices facilitate learning and technical expertise as well as shape participants' perceptions of the human body. By investigating the role that discourses, displays, and human bodies play in the training and socialization of medical students, T. Kenny Fountain contributes to our theoretical and practical understanding of the social factors that make rhetoric possible and material in technical domains. Thus, the book also explains how these displays, discourses, and practices lead to the trained perspective necessary for expertise. This trained vision is constructed over time through what Fountain terms *embodied rhetorical action*, an intertwining of body-object-environment that is the basis for all scientific, medical, and technical work.

This book will be valuable for graduate and advanced undergraduate courses in technical and professional communication (technical communication theory and practice, visual or multimodal communication, medical technical communication) and rhetorical studies, including visual or multimodal rhetoric, rhetoric of science, medical rhetoric, material rhetoric and embodiment, and ethnographic approaches to rhetoric.

T. Kenny Fountain is an assistant professor in the Department of English at Case Western Reserve University in Cleveland, Ohio. He received his PhD in Rhetoric and Scientific and Technical Communication from the University of Minnesota in 2008. He is a former Writing Center Assistant Director at Yeshiva College in New York City and a former Director of Writing Across the Curriculum at Bilkent University in Ankara, Turkey. His research interests include the rhetoric of science and medicine, visual studies of science, history and theory of rhetoric, communication in the disciplines, and theories of the body and embodiment. He has published in the journals *Medicine Studies* and the *Journal of Technical Writing and Communication* as well as the edited collections *Solving Problems in Technical Communication* and *Pluralizing Plagiarism*.

ATTW Book Series in Technical and Professional Communication

Jo Mackiewicz, Series Editor

For additional information on this series please visit www.attw.org/publica tions/book-series, and for information on other Routledge titles visit www.rout ledge.com

RHETORIC IN THE FLESH

Trained Vision, Technical Expertise, and the Gross Anatomy Lab

By T. Kenny Fountain

 Routledge
Taylor & Francis Group

NEW YORK AND LONDON

First published 2014
by Routledge
711 Third Avenue, New York, NY 10017
Simultaneously published in the UK

by Routledge
2 Park Square, Milton Park, Abingdon, Oxon OX14 4RN

Routledge is an imprint of the Taylor & Francis Group, an informa business

Library of Congress Cataloging-in-Publication Data

Fountain, T. Kenny.
Rhetoric in the flesh : trained vision, technical expertise, and the gross
    anatomy lab / by T. Kenny Fountain.
        pages cm. — (Attw series in technical and professional communication)
    Includes bibliographical references and index.
    1. Medicine—Study and teaching.   2. Medical students—Training of.
3. Rhetoric.   4. Communication in medicine.   I. Title.
R737.F66   2014
610.76—dc23   2014013510

ISBN: 978-0-415-74103-3 (hbk)
ISBN: 978-0-415-74102-6 (pbk)
ISBN: 978-1-315-81545-9 (ebk)

Typeset in Minion
by Apex CoVantage, LLC

CONTENTS

FIGURES

TABLES

SERIES EDITOR FOREWORD

We continue the ATTW Book Series in Technical and Professional Communication (TPC) with T. Kenny Fountain's *Rhetoric in the Flesh: Trained Vision, Technical Expertise, and the Gross Anatomy Lab*. The second book in the series, *Rhetoric in the Flesh* follows Liza Potts's *Social Media in Disaster Response: How Experience Architects Can Build for Participation*. Like Potts's book, Fountain's constructs what will become a long line of research-driven, useful, usable and—critically important—interesting texts.

Like Potts's book, Fountain's book reveals the diversity of research and practice related to TPC. Potts's book examines the role of social media—including its on-the-fly repurposing of content—in human-made and natural disasters; it rests in the overlap of crisis communication and TPC. Fountain's book examines the activities and discourses of the gross lab—the ways the participants "come to see the physical, material body as the anatomical body of medical knowledge" (22–23). Fountain's book, then, rests in the overlap of rhetoric of science and TPC.

As I noted in the foreword to Potts's book, developing books that are "research-driven, useful, usable, and interesting" means making abstract theory and complicated research comprehensible and making implications for TPC practice explicit. As you will see as you move through the chapters of *Rhetoric in the Flesh*, Fountain's book is certainly research driven, useful, and usable. But I have to admit that what first hooked me when Fountain came to me with his book proposal was the incredible interest of the subject matter: "how bodies, multimodal objects, and discourses coalesce" to form an assemblage that "structures human experience" for the participants who learn and teach in the gross lab (20). Past that initial hook, what interested me was the way that Fountain so readily connects the embodied rhetoric at work in the gross lab with the concerns and practice of TPC.

I am proud to have this book as the second in our series, and I am looking forward to the others that will follow.

Jo Mackiewicz
Editor, ATTW Book Series in
Technical and Professional Communication
3 September 2013

ACKNOWLEDGMENTS

This project began as a very different dissertation that I have rewritten a number of times to create this book. As such, my first debt of gratitude is to my committee. I thank my adviser Mary Lay Schuster. Her consistent encouragement, sound guidance, and quick response time steered me through my PhD program. The freedom she granted me to find my own way—all the while holding me accountable to solid research standards—made all the difference. Carol Berkenkotter offered needed structure and support at the beginning of the project, shaping my preliminary ideas and introducing me to research that is foundational to this book. It was her suggestion to consider the cognitive aspects of situated practice that eventually led me to theories of enaction. At the prospectus stage, Lee-Ann Kastman Breuch reminded me not to forget that I was studying a pedagogical space, a point that proved invaluable. She also offered sage advice and a sympathetic ear while I was putting together the book proposal. Jane Blocker's research, teaching style, and way of thinking about images, objects, and bodies influence and inspire me still. If anyone can be understood as my collaborator in this work it is Anthony Weinhaus, who was perhaps my biggest supporter. In the labs, he made me feel at home, and his openness, kindness, and willingness to lend a hand carried me at times when my own energy did not. Richard Graff, Bernadette Longo, Karen-Sue Taussig, and Art Walzer deserve praise for contributing so much to my intellectual and professional development.

Many friends listened tirelessly as I prattled on about bodies and cadavers, talk that shaped an earlier version of this project. I am thinking of Paul Anheier, Anthony Arrigo, Marnie Gamble, Elizabeth Kalbfleisch, Krista Kennedy, Dave Kmiec, Salma Monani, Zoe Nyssa, Amy Propen, Merry Rendahl, Greg Schneider-Bateman, Erin Wais-Hennen, and Jeff Ward as well as Michelle Anderson, Nancy

Antenucci, Anna Batsakes, Anna Martinson, Jenny Ovadia, Steve Ovadia, and especially Andrej Peterka.

I thank David Lee, Angela McArthur, Kenneth Roberts, and again Anthony Weinhaus, who believed in the study enough to grant me access to the labs, sent all the material I requested, promoted my work (to their students and colleagues), and listened to preliminary discussions and reports of my findings. Hundreds of participants—students, TAs, and instructors—were gracious and kind enough to let me hang out in the background with my notebook; they welcomed me and included me in the life of the labs. My hope is that I have represented their words, ideas, and activities with respect and accuracy. As well, I owe a debt to all those who gifted their cadavers to the labs. They gave their bodies not only for medical education but also my own.

Ellen Barton, Loel Kim, Jason Swarts, and Susan Wells all believed in my work at crucial moments when I needed it. I am thankful for the camaraderie of the CCCC Medical Rhetoric SIG, especially Barbara Heifferon. William Siebenschuh and Mary Grimm were kind and supportive department chairs who helped me find time to work on the book. Hannah Rankin was an excellent research assistant who assisted with transcriptions. Jonathan Scott Weedon's astute observations on processes of attention challenged me to consider cognitive models. Mary Assad's and Gregory Summer's enthusiasm for Burke has encouraged my own; their influence shaped my book's argument. The excitement of the freshmen in my Science and Culture of Anatomy course infused much life into what felt at times like a lifeless revision process. In significant ways, this book took shape through frequent conversations with Kimberly Emmons and Laura Hengehold, who helped me think through half-formed thoughts. In fact, the book would not exist had it not been for colleagues that championed my work, read proposals and drafts, gave their time and energy, or helped me guard my own time. For that I thank Kimberly Emmons, Christopher Flint, Kurt Koenigsberger, Robert Spadoni, and Athena Vrettos.

This research was funded in part by a thesis grant and a doctoral dissertation fellowship from the Graduate School of the University of Minnesota and an ACES+ grant from Case Western Reserve University. For assistance with image permissions, I thank Georgina Gomez (Gunther von Hagens's *Bodyworlds* and the Institute for Plastination), Niki Russell (University of Glasgow Library, Special Collections), and Michael Sappol (National Library of Medicine). I presented sections of this book at academic conferences and an invited talk at Brock University in St. Catherines, Canada. I thank those audiences for the questions they asked. In 2010, a different version of chapter 4 appeared as "Anatomy and the Observational-Embodied Look" in *Medicine Studies*.

Two anonymous reviewers provided insightful and generous feedback that shaped the final project. At Routledge, Linda Bathgate supported the book and competently guided it through the process with kindness and a great sense of humor. Jo Mackiewicz, the ATTW series editor, provided limitless patience, a keen eye, smart suggestions, and tireless support of the project and me. I also thank Jo's editorial assistants—Kelcie Sharp, Allie Martin, and Mallory Porch—for their hard work on the manuscript. To my partner Ryan Sherman, no thank you will ever be enough. I am unbelievably grateful for his patience, understanding, twisted wit, and most especially his calming presence and his love. I have promised him it will not always be this crazy, and I expect him to hold me to it. Lastly, I thank my family back in Georgia. I'm sorry for the long absence finishing this book caused.

This book is dedicated to my mom and dad—Mary Alice and Bun—a nurse and a carpenter, whose firsthand experiences of medicine and embodied skill somehow prepared me to write this.

1

INTRODUCTION

Developing Expertise and Learning to See

Technical communication, as an academic discipline and as a field of practice, has grappled with questions of expertise for centuries. Most historical accounts of the profession, such as Connors's (1982), Longo's (2000), and Kynell-Hunt and Savage's (2003), make visible our persistent struggles over authority, legitimacy, and the right to claim expertise. Perhaps this focus on disputes of power is not surprising; after all, claims of expertise, as Hartelius (2010) reminds us, entail rhetorical contests over "ownership and legitimacy" between "autonomous" (or genuine) expertise and "attributed" expertise (a type that gains its credibility from others' recognition) (3–4). In technical and professional communication (TPC), questions of expertise—autonomous or attributed—are structured by what Geisler (1994) describes as a "dual problem space framework" that divides expertise into two distinct components: a content domain and a rhetorical process domain (82). Various stakeholders in the academy and the professions use this division to grant expertise to some while withholding it from others (89). In TPC, those with more social, political, and economic capital apply these alleged distinctions between content and form as well as between subject-matter and rhetoric to identify those who do the supposedly real work of science (scientists) and those who merely write it up (writers) (Longo 2000). Following this bifurcated approach, we inevitably (perhaps inadvertently) split knowledge from know-how, or what Ryle (1949) famously termed "knowing that" and "knowing what" (28). In adhering to this binary that Carter (2007) calls "conceptual knowledge" versus "process knowledge" (387), we separate knowledge of a domain from the ability to communicate that knowledge.

While some people do possess greater expertise in areas of content and some in areas of communication, we should not view communicative expertise, or the expertise involved in communication, as easily divisible from what I call technical expertise, or the specialized knowledge that professionals use in technical domains, including science, medicine, engineering, math, architecture, and fields intimately connected with them, like TPC. As Geisler (1994) illustrates, communicative expertise requires not only subject-matter knowledge and rhetorical skills, but also involves ways of thinking and problem solving coupled with the ability to understand the rhetorical and social aspects of communication (92). Similarly, Tardy's (2009) exploration of communicative expertise complicates these dichotomies by demonstrating how formal, process, and rhetorical knowledge of a content domain depend on both knowledge and know-how. This dualistic model stems, I contend, from a view of knowledge as an object to possess, a focus on traditional written texts, and a disembodied approach to meaning making in general. When we consider expertise a form of activity, as Regli (1999) encourages us to do, we find that expertise, particularly technical expertise, cuts across these dichotomies. Being an expert involves all of these allegedly disparate areas: knowledge and know-how, communicative abilities and subject-matter knowledge, even autonomous and attributed expertise. Technical expertise, as I demonstrate, is not about having and not having, but about seeing and doing—specifically, using the body's capabilities to develop and enact knowledge and know-how. Technical expertise, as such, is a type of trained vision we acquire through embodied practice.

Rhetoric in the Flesh is a book about expertise and vision; more precisely, it is about learning to see in technical domains. A few conceptual questions motivate my inquiry: How do newcomers to a professional group come to see according to the practices, values, and beliefs of that community? And what part do images and visual displays play? To move beyond an exclusive focus on conventional word-based documents, Rhetoric in the Flesh attends to the ways visual images and objects participate in processes of education and socialization. Specifically, I describe how we acquire technical and communicative expertise by learning to see as a member of a particular group. Through an ethnographic analysis of one medical school's gross anatomy program, I explore how participants develop knowledge and expertise in a scientific field, in this case, cadaver-based anatomy. In telling this story of images and expertise, I explore the mutually structuring relationship between visual displays and bodies in the gross lab, and I do so in order to interrogate the rhetorical function of images and objects in the production of medical knowledge and the development of medical professionals.

Yet, images are not the only means of producing this skilled perspective. Participants in this domain of technical work also learn to see and recognize the anatomical body through rhetorical discourses that introduce students to the values, perspectives, and beliefs of medicine. The development of anatomical knowledge and know-how, including a medical professional's clinical detachment, coincides with the skillful use and suasory force of visual and rhetorical displays of evidence and values, or what I describe as apodeictic and epideictic demonstrations. By focusing on the contemporary anatomy lab as an assemblage of multimodal displays, medical and institutional discourses, and rhetorical and embodied practices, I explain how over time participants develop a perspective that shapes how they view and even respond to the human body. Thus, as I demonstrate, technical experience and professional dispositions are, in a significant way, rhetorically trained. Furthermore, participants develop this expertise, this trained perspective, by learning to engage bodily with the displays, objects, and discourses of this technical space.

Expertise and Trained Vision

Geisler (1994) defines expertise as a type of knowledge that "goes beyond everyday understanding," and which a person acquires through "specialized training and practice" (53). She contends that expertise is transmitted most often in formal educational settings through "interactions over texts" (54). Literate practices, namely reading and composing written texts, are crucial to gaining, communicating, and transferring expertise (92). Most TPC studies of expertise acquisition have focused on these conventional documents, such as reports, manuals, and the academic and professional writing of novices. This concern with the written word is unsurprising given the cultural power and semiotic complexity of writing particular to a field that began with "technical writers" and "teachers of technical writing" (Connors 1982; Kynell and Tebeaux 2009). However, as research has shown, knowledge production and expertise transmission in TPC are not exclusively instantiated through word-dominant texts. As Sauer (2003) and Johnson (2006) demonstrate, nonwritten modes, from gestural to oral communication, often convey knowledge and know-how necessary to the work of science and technology.

One prime example of a non-word-dominant mode of TPC that has received increasing attention is the image-based text. In scientific, medical, and technical professions, images play a major role in constructing expertise, making arguments, and transforming the perceptions of audiences both inside and outside of science (Donnell 2005; Gross, Harmon, and Reidy 2002). The production, interpretation, communication, and distribution of scientific knowledge require a host of visual representations deployed or, as Latour (1990) asserts, "mobilized" to make and communicate knowledge claims (40). These scientific visual displays are evidentiary artifacts generated to represent, contain, and shape research findings or observable phenomena (26). Participants in scientific, medical, and technical contexts accomplish their work and produce relevant phenomena through processes of verbal and visual inscription, activities that transform "material substance into a figure or diagram" that participants then use to back up or refute statements, hypotheses, and claims (Latour and Woolgar [1979] 1986, 51). In no small part, these displays constitute scientific knowledge by making it visible.

With advances in technology, visual displays—and even documents once understood as primarily word based—are increasingly multimodal objects that communicate through various channels, discourses, and media (Kress and van Leeuwen 2001, 2–4). As a result, many researchers have criticized as unhelpful and inaccurate once-rigid distinctions between visual and verbal texts (Kostelnick and Hassett 2003; Kress and van Leeuwen 2001; Wysocki 2004). In scientific and technical contexts, displays of all kinds are multimodal objects that signify through a combination of visual, verbal, spatial, and even physical or material modes. These multimodal displays of scientific research—such as graphs, charts, photos, and 3-D representational objects—carry a persuasive and ontological force that influences the formation and dissemination of scientific arguments, a force that shapes how scientists and nonscientists alike view these objects and phenomena. Multimodal displays, then, not only communicate knowledge and know-how; they also train participants to view this knowledge and the objects of their profession in a certain light. As such, technical expertise—an advanced mastery of the knowledge and know-how of a technical domain—involves learning to see and use objects, tools, and discourses according to the authorized practices of a particular group.

In this way, the visual practices of a community enact a type of socialization, developing in participants what Goodwin (1994) describes as "professional vision," or "socially organized ways of seeing." (606) This socialization is at once a philosophical worldview, a disciplinary framework, and a visual orientation allowing community members to see social phenomena (tools and objects rendered by a social group) according to the logic of that profession. For example, an archeologist-in-training develops the professional vision of her field by learning to make scientifically relevant observations in "a patch of dirt" by reading it through the interpretative frame provided by visual displays, such as the Munsell color chart (606). By learning how to use multimodal displays as interpretative frames, or what Goodwin (2000) terms an "architecture for perception," participants come to see the objects of their environment according to the culture of that group (34). To develop this trained perspective, participants must learn to use and read these displays correctly, a skill they learn through immersion and socialization into the literate, visual, discursive, and embodied practices of a group (Goodwin 2000).

Scientific and technical practices produce multimodal displays and objects plus ways of viewing those objects that shape the perceptions and perspectives of participants. This trained perspective, for Grasseni (2007), is the result of a participant's embodied interactions with a "richly textured environment" of objects, bodies, and images (207). In her ethnographic study of the training of cattle breeders, Grasseni redefines vision as "the capacities and capabilities" derived through apprenticeship and training in a particular community of practice (215). That is, "skilled vision" is cultivated when we gain the ability to see by way of the interpretative frames and objects situated in "an ecology of practice" (Grasseni 2004, 41). Through the development of this skilled perspective, a person learns to "see as" a member of a group in a way that makes expertise material and visible to members and nonmembers alike (Grasseni 2007, 7). Skilled vision not only shapes and communicates meaning; it also fundamentally makes possible the meanings, attitudes, and practices of a group (5).

Situated practice, then, not only produces knowledge and know-how, but also ways of viewing that knowledge and the objects and discourses of that knowledge, ways of being in the world that shape the lived experience of participants. This "trained vision," as I call it, is the organization of perception through the interplay of multimodal displays and objects, interpretative frameworks, and the situated activities that create and deploy them. Trained vision, which we learn through formal training and informal socialization, originates with the material practices of a group and our embodied participation in those practices. We develop expertise, technical or otherwise, not just through socialization into a community's discourses, genres, documents, multimodal displays, and objects. Instead, we develop expertise when we develop the skilled capacities necessary to use the discourses and objects, the displays and documents, according to the explicit and tacit rules of that community. Acquiring these capacities with objects and discourses involves the acquisition of particular ways of seeing, thinking, and being in the world. This trained vision both coincides and coconstructs expertise. Thus, the literate activities and bodily capacities that result in technical expertise originate with our embodied interactions with the discourses and multimodal objects of a community.

Clearly then, expertise involves more than learning the knowledge of a group; it involves learning to perform tasks as a member of that group as well as gaining the social role that allows one to perform those tasks. The more I learn to perform the practices of a domain, the more mastery I attain. Becoming a scientist, doctor, engineer, or a technical communicator requires a person learn content knowledge and rhetorical knowledge, but not always as separate forms of knowledge. Through embodied socialization into the language, genres, discourses, documents, displays, and practices of a technical domain, we develop technical expertise. To learn these ways of knowing, seeing, and being, we must immerse ourselves in practices of that group. Only then can we acquire the knowledge and develop the skills and dispositions of that community. Learning to see means learning to recognize, appreciate, and understand the images and objects of our discipline; it also means learning to share the same values and convictions prized by that discipline.

The Focus of the Book

The anatomy lab offers a clear vantage point to view how multimodal objects, together with rhetorical discourses, form the trained vision of technical expertise. My analysis of these embodied and rhetorical processes develops from my ethnographic observations and immersion in one American medical school's human anatomy program, specifically, the two gross anatomy courses that composed that program: (1) ANAT 600, a dissection lab for first-year medical and dental students using fresh cadavers; and (2) ANAT 303, an undergraduate course using prosected (or previously dissected) cadavers. Through an analysis of field notes, interviews, course material, and photographic documentation I collected during my fieldwork, I demonstrate how anatomy education is a social, embodied, and deeply rhetorical endeavor. That is, these gross anatomy labs use two types of rhetorical demonstrations: (1) multimodal displays used to exhibit anatomical knowledge; and (2) institutional discourses used to represent and impart knowledge and, more importantly, to inculcate the values of anatomical study and medical science. These demonstrations of evidence and values correspond broadly to ancient rhetorical notions of *apodeixis* (displays of proof) and *epideixis* (displays of cultural values). Together, the apodeictic displays, in the form of multimodal objects and rhetorical demonstrations, and the epideictic verbal discourses and narratives induce in participants a trained vision that shapes how they view the anatomical body of medicine and the lived body of human experience.

This trained vision is the result of participants' bodily encounters with multimodal objects—such as images, photographs, x-rays, and cadavers—and their repeated exposure to the epideictic discourses and processes that laud cadavers as crucial to learning and eulogize cadaveric donors as altruistic gift givers. To learn, teach, and communicate anatomical knowledge, participants must recognize the anatomical body in the multimodal displays. They accomplish such recognition by learning how to view, touch, move, and manipulate the bodies of the lab—the nonhuman displays that represent anatomy and those living and dead bodies that present anatomical structures. Thus, the multimodal displays and embodied practices of this medical setting facilitate learning and technical expertise while shaping participants' perceptions of the body. Also, the rhetorical discourses that influence how participants work with and feel about these objects fundamentally shape participants' abilities to engage in the embodied activities of demonstrating, observing, and dissecting. By investigating the role that discourses, multimodal displays, and human bodies play in the training and socialization of medical students, I explicate how these displays, discourses, documents, and practices lead to the trained vision necessary for expertise. I do so by illustrating how embodiment and bodily practices are paramount to any configuration of rhetorical action and TPC.

By explicating the embodied practices and nonscientific rhetorics that enact the trained vision of medical professionals, *Rhetoric in the Flesh* contributes to our knowledge of how learning to see—in any technical field—develops through formal training and informal socialization. Over time, a participant develops trained vision through what I call the "embodied rhetorical actions" of a domain, actions that make possible all scientific, medical, and technical practice, knowledge, and expertise. As such, there can be no technical and professional communication without the embodied practices necessary to read, compose, and distribute rhetorical displays, documents, and discourses. Further, we can only acquire and communicate expertise through our embodied interactions with the rhetorical displays that constitute our domains of practice.

Training Vision in a Gross Anatomy Lab

Since the late middle ages, the study of cadaver-based anatomy has served as a primary introduction to anatomical knowledge and as the foundation of medical education—the rite through which students not only acquire seemingly authoritative knowledge of the body but also become practitioners of medicine (Carlino [1994] 1999; Park 2006). The practices of dissection, demonstration, and observation communicate this authoritative knowledge to those who seek to reveal the body's supposed secrets. For centuries, the activities of cutting, presenting, and viewing human bodies have dominated medical education in America (Sappol 2002; Wells 2001), England (Richardson 2000b), and Europe (Carlino [1994] 1999; Cunningham 2010; French 1999; Klestinec 2011). As a contemporary medical discipline, human anatomy is primarily the study of the structure of the body (Moore and Dalley 2005). As the foundational language of medicine, anatomy is not just a list of parts to be identified but a complex knowledge system using mostly Latin and Greek terms, anatomical planes such as median, sagittal, coronal, and transverse, as well as spatial orientations of comparison and relation such as superficial and deep, medial and lateral, and posterior and anterior (Moore and Dalley 2005, 5–8).

But instead of simply observing and memorizing anatomical terms and structures relating to the whole of the human form, students must learn those structures systemically, learning individual parts as components of larger systems that work together to create the functioning human. For more than four centuries, students and teachers have used visual illustrations and other images to aid them in dissecting, identifying, teaching, and learning (Carlino [1994] 1999; van Dijck 2005). These displays depict anatomical knowledge through a complex network of multimodal objects—from dead and living human bodies, x-rays, CT (computed-tomographic) images, plastic models, and digital imaging, to images from atlases (bound collections of scientific images designed to train novices). Today, as it has been since the sixteenth century, students and even teaching assistants (TAs) learn anatomy primarily through their skillful interactions with these multimodal displays and the physical human bodies they represent. Take ' for example, the photograph in figure 1.1.

Though we cannot see them here, there are, of course, cadavers, either lying atop tanks that look like tables or tucked away inside them. What we do see are the first bodies students encounter, the 2-D and 3-D representations that cover the walls and whiteboards, the ones that rest on tables and countertops.

These photographs, x-rays, color illustrations, line drawings, and professionally produced posters, not to mention the cadavers, constitute the primary texts

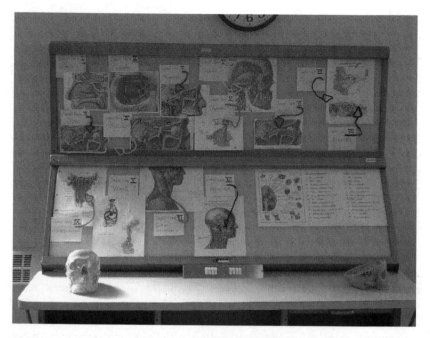

FIGURE 1.1 An x-ray light box with images and plastic skulls.

and displays of the course, the tools necessary for learning, teaching, and enacting the anatomical body. Through their bodily interactions with these displays, participants make anatomical knowledge discernable on and through these visuals. In the process, students and TAs develop a trained vision—in this case, an anatomical perspective—that shapes their perceptions of the human body, both the dead bodies on the tables and the living bodies that move among them.

But learning to see in anatomy labs, as in life, requires more than merely looking and knowing. Vision, as a social activity, requires other bodily movements and perceptual orientations. The famous nineteenth-century anatomist Luther Holden (1887) described anatomy laboratory learning as a set of practices that involve the training of the entire body: "Our main object, therefore, is to induce in students the habit of looking at the living body with anatomical eyes, and with eyes, too, at their finger ends" (1,025). Through the habitual examination of bodies (both living and dead), students develop "anatomical eyes" and "surgical fingers," perceptual tools that offer a way of conceiving of the body. This dispositional tendency originates with the lab's bodily practices. The kinesthetic and the tactile experiences the cadaveric specimen affords are indispensable for recognizing and enacting the anatomical body.

Many anatomists view the development of manual dexterity as a major goal of the gross lab (Ellis 2001, 149). For Ellis, dissection is the only "educational modality in the preclinical" curriculum that teaches students "how to use their hands," not only how to hold cutting instruments but also "how to appreciate tissues" (149). Hanna and Freeston (2002) praise the physical, three-dimensionality of cadaveric anatomy because it offers students manual dexterity and the appreciation of "tissue planes and the scale and depth" of anatomical structures (377). The word "appreciation" implies that students come to an awareness of, admiration for, and even a pleasure in, the materiality of the human body not just by studying it or cutting it but by touching it with their hands. Take for example the technique known as "blunt dissection," whereby a dissector inserts her hand (positioned in what looks like a karate-chop gesture) into the cadaver to divide and explore the body's anatomical planes. By sliding a gloved hand between the skin and the muscle beneath, a student can dissect that region and get a tactile awareness of depth, firmness, shape and, to some extent, texture.

To make sense of the unruly cadaveric body, students learn to transform haptic (touch-based) sensations of depth, texture, and pressure into information about the body (Prentice 2013). One reliable way to differentiate nerves, veins, and arteries in the cadaveric body is not through visual recognition but instead through haptic recognition. A body's nerves, veins, and arteries do not, after all, appear color-coded as white, blue, and red (respectively) as they do in textbook illustrations. Even with gloved hands, as all hands are in the lab, students can easily feel differences once they recognize them. More than just a supplement to seeing, touching becomes a type of perceptual tool that extends the capacities of seeing. Observation in the gross lab involves more than simply looking; in order to recognize and learn anatomy, participants must connect evidence gained from both vision and touch.

Learning medicine, then, requires more than just learning medical knowledge; it requires learning to use that knowledge to see, feel, and think the way a physician does, and learning to respond to displays and objects, such as the body, as a physician would. This professional practice, beginning in the gross lab, trains participants to view bodies by way of anatomical and medical knowledge as well as the physical practices and skilled capacities of the medical profession. More than just a class to learn anatomy (a pedagogical space) or a centuries-old medical initiation ritual (a professional space), today's gross lab is a space of contradiction—of creation through destruction, visibility through disappearance, beautiful order through gruesome chaos, and appreciation for the body through what no doubt seems like disrespect. The lab is also a space of *poiesis* (making). What is being made is not simply a medical professional or a specialized knowledge of the body, but a type of body—namely the anatomical body—and a way of seeing, which I call anatomical vision. In this often-misunderstood setting, future medical professionals learn to engage their entire bodies in order to recognize, teach, debate, and communicate anatomical knowledge, all while they develop medical expertise and a powerful anatomical vision.

Field Site and Methods

My ethnographic data derives from fieldwork I conducted over the course of a year at one large medical school's human anatomy program. I conducted this research with IRB approval and the consent of the anatomy program. All names of participants are pseudonyms. At that time, the program housed two large-enrollment laboratory courses: ANAT 600, which I observed in the fall of 2006, and ANAT 303, which I observed in the spring of 2007. The medical and dental gross anatomy course, ANAT 600, was an eight-week, intensive dissection lab complemented by a conventional lecture that covered gross anatomy (anatomy detectable with only the human eye) and introduced students to radiographic (x-ray), histological (microscopic), and embryological anatomy. At the beginning of the course (taught only in the fall), the program administrators divided students into groups of four, who became "body buddies." Over the course of the term, each team completely and collaboratively dissected the one human cadaver assigned to it. To assist in the dissections and demonstrations (the actual teaching of anatomy), 15 third-year medical students were assigned as TAs to aid the four instructors of the course.

Offered each spring, the undergraduate course ANAT 302 was a massive multisection laboratory class serving more than 400 students from various biological, medical, and health-related majors. This lab, in conjunction with the lecture (ANAT 301), provided a general overview of gross, histological, and radiographic anatomy. Unlike the medical and dental school course, this course was a prosection lab, meaning that students enrolled in the course did not usually perform dissections themselves but instead learned through observation and demonstration from already dissected cadavers. The TAs, supervised by the director of the undergraduate lab, did the actual dissections and laboratory preparations each Friday afternoon. The goal of these "TA prep labs" was to create an educational and representational environment where undergraduates could learn human anatomy. The TAs were a select group of advanced undergraduates who, after passing the course with an A, formally applied for the position. Thus, these TAs assumed explicitly pedagogical and professional roles often granted only to graduate students.

I chose to blend observations and in-depth interviews with rhetorical analysis of displays, objects, documents, and discourses in order to understand the practices that make the anatomy courses possible, practices that enact and instantiate the complex socialization and formation of perception that anatomical education entails. My collection of qualitative and interpretative data involved four components: (1) direct observations of actual laboratory sessions and the lab preparation meetings for both courses; (2) audio-recorded interviews with students, TAs, and instructors from the undergraduate and medical/dental courses as well as staff of the anatomical bequest program who procure and prepare the bodies and work with donor families; (3) teaching material from both courses, including course websites, handouts, lab notebooks, and textbooks; and (4) photographic documentation of the spaces and objects over the course of my fieldwork. My approach to data collection and analysis was similar to common grounded theory strategies discussed by Charmaz and Mitchell (2001): (1) the simultaneous collection and analysis of data to uncover "emergent themes" in the early data; (2) a concentration on the social and cultural processes found within the data; (3) "an inductive construction" of categories that "explain and synthesize these processes"; and (4) the "integration of categories into a theoretical framework" that explains these processes (160). (For a fuller discussion of methods, see Appendix: Data Collection and Analysis.)

I analyzed my typed field notes, interview transcripts, and other documents I collected using an analytic coding method advocated by Emerson, Fretz, and Shaw (1995, 144). Specifically, I developed codes and categories directly from the data through an iterative process of open coding, focused coding, and memo writing that involved reading, rereading, analyzing, and returning again and again to my field site (Emerson, Fretz, and Shaw 1995, 144–45; Charmaz and Mitchell 2001, 165). The following research questions guided my reading and interrogation of data: (1) What is the role of the body, both living and dead, in teaching, learning, communicating, and confirming anatomy in a cadaver-based laboratory setting? (2) Through which modes, media, and technologies is the body displayed? (3) In what material practices do lab participants engage? (4) What roles do the discourses, multimodal displays, physical objects, and material practices play in the work of the labs? (5) How do these discourses, objects, and practices influence how participants communicate, interact with, and view the human body? My goal was to uncover how anatomists and students visually, rhetorically, socially, and materially enact the anatomical body as well as how this enactment, this making, affects those same participants.

Thus, my initial interests centered on the visual and material rhetoric of this medical and educational space. But as I began to interact physically with the lab's visuals—the bodies, atlas images, even wall posters—I came to realize the importance of embodiment and embodied practices. I remember the particular afternoon when I decided to participate in a practice exam. With notebook in hand, I found myself standing incredulous above an open torso, struggling to identify an anatomical structure tagged with bright red string. I reached in and touched what I thought must be a nerve. As soon as I felt its thin springiness and slight elasticity between by gloved fingers, I knew at once I was correct. And considering the depth inside the body my hand had reached, I had a good idea which nerve it was. At that point, I realized I knew the structure because I knew its name, basic function, and location, all of which I was able to accomplish because I could confirm this knowledge through my own bodily perceptions—sight, touch, movement, depth, and texture. My realization that embodiment played a prominent role was confirmed over and again through my observations, interviews with participants, analysis of visual and verbal data, and personal experiences of studying and learning anatomy. To make sense of these anatomical practices that now seemed unmistakably and simultaneously rhetorical and embodied, I sought out existing theoretical models to help me make sense of my data and my experiences.

Most of us will agree that research, scientific or otherwise, is never neutral and that to observe a phenomenon is both to intervene in it (Barad 2007; Hacking 1983) and to view it through a lens of unavoidable theoretical assumptions (Burke [1935] 1984; Hanson 1958). Even still, an explicit use of theory in qualitative research can be a complicated and controversial choice. While most grounded theory and ethnographic approaches are adamant that theoretical concepts "arise from analyzing data," this outlook need not preclude the use of existing theories to make sense of that data, particularly if those theories "sensitize" researchers to processes observed in the data (Charmaz and Mitchell 2001, 169). Take for instance Denzin's (1978) model of triangulation, understood as the use of multiple methods to support and strengthen one another as well as corroborate findings. Denzin (1978, 2007) includes not just methodological triangulation but also theoretical triangulation (the incorporation of multiple theoretical perspectives) as a way to provide the researcher an in-depth understanding of the complex processes involved in any social phenomenon.

Yet to accurately understand and perhaps even faithfully capture the practices and objects we investigate, researchers must be willing to adapt as well as adopt theoretical models by putting them in conversation with empirical data. In writing *Rhetoric in the Flesh*, I turned to rhetorical theory (my disciplinary foundation) and phenomenological and cognitive theories of embodiment. I found that these highly compatible approaches resonated with my data, and also supported and enlivened each other, offering a perspective that each alone arguably lacks. As I will demonstrate, classically inspired rhetorical theory, together with phenomenological and cognitive theories of embodied action, offers a robust conception of human activity. In other words, inside and outside technical workplaces, rhetoric is dependent on our bodily capacities for action and perception, just as these bodily capacities and activities are caught up in larger rhetorical processes.

Rhetoric and Embodied Action

Rhetoric involves, in Mailloux's (2006) words, a "double nature" as a productive and analytic art (38). In its productive mode (*rhetorica utens*), rhetoric entails the creation of a persuasive force by way of some text, speech, image, or object (Burke [1950] 1969, 36). Here rhetorica utens includes both the guidelines for producing persuasion and the actual production of a rhetorical text or object. In its receptive or analytical mode (*rhetorica docens*), rhetoric is the study or analysis of rhetorical texts and objects (36). According to Walker (2011), rhetorica docens also implies a theory or explanation of the practices that constitute persuasion (1). Burke ([1950] 1969) defines rhetoric (in the form of rhetorica utens) as the use of symbols by humans "to form attitudes and to induce actions" in others (41). For Burke, this inducement to action and cooperation requires a process of identification on the part of the person or social group to whom this suasory force is directed (22). The rhetorical force of language, discourse, and physical objects encourages certain actions and perceptions in social actors and in the process builds social bonds among those actors moved by a particular message or worldview. For Burke, rhetorical force is an "ingredient" in all forms of socialization, in subtle and not so subtle ways (39). As such, rhetorical analysis involves an investigation of how discourses, texts, and objects (both human and nonhuman) work to constitutive even generate social phenomena by inducing actions and beliefs in others.

I define rhetoric (or rhetorica utens) as the use or generation of displays, texts, discourses, images, or objects to exert a persuasive and ontological force that forms attitudes and beliefs, encourages and structures actions and practices, and as a result coconstitutes and coenacts social phenomena. Rhetoric involves traditional and nontraditional texts and objects (including oral, written, visual, gestural, and multimodal texts) as well as intentional and unintentional audiences and effects. I take rhetorica docens to be a mode of analysis or a capacity for observation understood as a type of skilled perceptive or trained vision made possible by a person's (or a rhetor's) formal training and socialization. The persuasive and ontological force of language, images, and objects makes possible all rhetorical actions, scientific or otherwise, by authorizing certain meanings, messages, and perspectives while discouraging others. When we learn to view the world or its objects according to the logic of a participant group, we engage in rhetorical processes of meaning making that shape our embodied actions in the world and our views of the possible.

Following Haas and Witte's (2001) study of the role of embodiment in engineering writing, I use the term "embodied" to refer to actions that are "accomplished by means of the human body," usually involving "skillful and often internalized manipulation" of one's own body and the tools of the environment (416). By embodied practices, I mean types of habitual social action and social knowledge that are constructed and communicated in and through the materiality and physical movements of the body. Embodied practices are reminiscent of Bourdieu's (1980) concept of *habitus*, or the "acquired system of generative schemes" (55) or dispositional tendencies "constituted in practice" (52) that "generate and organize practices and representations" (53). By defining embodied practices as I do, I understand all human activities as embodied in that there are no practices without bodies. As Merleau-Ponty ([1945] 2005) has made plain, "we are our body" in that "we are in the world through our body" (239). Thus, we only know the world because we are embodied in the world, and that world comes to us through the bodily phenomenon of perception. The body, then, is more than just the material inscribed by discourses; it is our means of making sense of such discourse and our capacity for action.

Like Haas and Witte's (2001) work, Sauer's (2003) study of mining safety documentation demonstrates that technical communication involves the body. Yet much of the research in TPC and rhetorical studies still seems to rely on either disembodied computational models of cognition such as Hutchins's (1995) distributed cognition or poststructuralist theories of the body such as Foucault's ([1977] 1995). While these approaches are useful, they cannot account fully for the ways perception, cognition, and knowledge production depend on the lived human body, particularly in its relation to objects and discourses that surround it. My work addresses this gap by drawing from complementary theories of embodied cognition: the enactive model of mind (Noë 2006; Thompson 2010; Varela, Thompson, and Rosch 1991), the ecological approach to visual perception (Gibson 1986), and the skillful coping mode of learning (Dreyfus 2005).

Though I explore these theories in more detail in the chapters to come, a concise overview is in order. The enactive approach of Varela, Thompson, and Rosch (1991), which they view as a continuation of Merleau-Ponty's ([1945] 2005, 1968) phenomenology, asserts that the human mind is enacted through sensorimotor activity with the objects of our environment. Perception and cognition do not follow the computational model that reduces the human to a signal-processing machine that merely responds to environmental stimuli. Instead, perception and cognition occur as we explore our connections to the world through embodied actions and the exercise of bodily skills (Noë 2006; Thompson 2010; Varela, Thompson, and Rosch 1991). Perception is "a kind of skillful activity" accomplished by way of the body, not just the brain (Noë 2006, 2). We make meaning in the world through our bodily engagement with objects, tools, and displays that seem to function as corporeal extensions and incorporations. By engaging with these objects' affordances, or the opportunities for action they seem to make possible, we develop the skilled capacity to make meaning with those objects and to perform meaningful action in this ecology of perceptual possibilities (Gibson 1986). Learning, then, is a matter of developing the bodily skills and habits necessary to cope with the objects in one's environment and to do so in a way that becomes almost second nature (Dreyfus 2005).

In line with scholars in classical rhetoric (Fredal 2006; Hawhee 2004) and material rhetoric (Blair 1999; Propen 2012; Schuster 2006), my work illuminates how rhetoric, like cognition, always involves a great deal more than just the brain. I do this by explaining rhetorical action as an embodied process of making meaning, inducing action, and enacting social phenomena. Incorporating work in phenomenology and cognitive science, *Rhetoric in the Flesh* constructs a theory of embodied rhetorical action that explains how objects, bodies, and discourses together generate a professional (in this case, medical) subjectivity that emerges in practice and is rooted in bodily activities. Embodied rhetorical action is my term for the connection of objects, discourses, lived bodies, and embodied practices—an object-body-environment intertwining—that develops in participants the skilled vision that makes all technical and professional knowledge possible. As a theory relevant to other medical, scientific, technical, and workplace settings, embodied rhetorical action expands conventional notions of how bodies use and interact with visual objects, how discourses structure human activity, and how these physical and rhetorical engagements enact social phenomena and ways of being. I developed this approach to rhetoric, embodiment, and action through a dual process of reading my data in light of rhetorical theory and phenomenology as well as reading these extant theories in light of my empirical data. From this mutual interrogation, I move from observations of one professional setting to broader concepts applicable to other domains of situated activity. In a sense, I offer *Rhetoric in the Flesh* as an evidentiary demonstration of this theory-building process.

While my explicit engagement with phenomenology and cognitive science is perhaps novel to TPC and rhetorical studies, my interdisciplinary approach reflects what has become an integral feature of our field. Rhetoricians and TPC scholars concerned with the scientific and medical past have looked to historical methods (Bazerman 1988; Berkenkotter 2008; Campbell 1989; Longo 2000; Tebeaux 1997). Those interested in the present have turned to cultural studies (Scott 2003), qualitative methods of anthropology and sociology (Graves 2005; Lay 2000; Sauer 2003; Segal 2005; Spinuzzi 2003, 2008; Winsor 2003), and quantitative analysis (Condit 1999; Gross, Harman, and Reidy 2002). Segal (2005) observes that many researchers studying medicine and health from a variety of disciplines are not only doing rhetorical studies, but also understand their work specifically in those terms (313). She provides the example of Dumit's (2004) use of Burke in his ethnography of the creation and deployment of positron emission tomography (or PET scans). I could also point to Martin's (1991) work on the "sleeping metaphors" of stereotypical gender used in descriptions of sexual reproduction (501), Sharp's (2006) discussion of the rhetoric of organ transplantation, or Saunders's (2008) turn to classical rhetorical concepts in his ethnography of a CT imaging suite.

These and other contributions of "nonrhetoricians" to rhetorical studies are worthy of notice because many of rhetoric's theoretical sources, like Bourdieu, Foucault, and Latour, are drawn from other fields (Segal 2005, 313). Fuller (1997) even encourages rhetoricians of science to use methods that are "more interchangeable" with other fields of science studies, namely sociology. He maintains that relying less on classical sources will enable more interesting readings of the rhetorical and textual features of science (279). While I am a strong advocate of interdisciplinary inquiry into TPC and rhetorical studies of science, I see no need to discard classically influenced theories, especially when many of the classical formations of rhetorical practice have yet to be fully explored in TPC and the rhetoric of science. I am thinking in particular of the ways displays of evidence (apodeixis) work together with displays of cultural values (epideixis) to facilitate the work of science and medicine.

Technical Training through Rhetorical Displays

In *Rhetoric in the Flesh*, I contend that in spaces of technical training, in this case anatomy education, students learn through prolonged encounters with displays, documents, and discourses that function as apodeictic and epideictic demonstrations. In the original Greek, the terms *apodeiktikos* and *epideiktikos* derive from the same root, *deiktikos* or *deixis*, meaning "exhibit," "show forth," and "make known" (McKeon 1987, 19; Prelli 2006, 2); "display" and "demonstration" (Latour 2008, 445); as well as "point" and "indicate." Scholars credit the division between these two forms of display to Aristotle's *Rhetoric* (2007) and other works influenced by Plato (Latour 2008; McKeon 1987; Prelli 2006). McKeon (1987) argues that Aristotle used both terms to describe "processes of presentation and manifestation," but apodeictic was more specifically "proof" and epideictic specifically "display" (19). Apodeixis, which Kennedy (2007) characterizes as Aristotle's term for "logically valid, scientific demonstration," eventually came to signify demonstrations of evidence like those found in mathematical proofs (33n20). But, as Netz's (2003) history of Greek mathematics explains, this formation of apodeixis as scientific evidence developed in part from the rhetorical notion of epideixis, specifically the oratory culture of ancient Greece (293).

For the Greeks, epideictic performances were largely speeches delivered outside judicial and political contexts, including funeral orations (*epitaphios logos*), ceremonial speeches (such as *enkomion*), and literary performances (such as *panegryikos*) (Timmerman and Schiappa 2010, 12; Walker 2000, 7). Offering perhaps the first systematic definition, Aristotle, in book 1, chapter 3 of the *Rhetoric* (2007), conceived of the epideictic genre as ceremonial speeches of praise and blame meant to explicate or justify what was honorable or shameful by exhorting or criticizing some person, object, or concept (Kennedy 2007, 49). Usually through amplification, these speeches measured the worth of (for example) a person according to what was considered virtuous action (76). Aristotle's formation of epideixis as primarily ceremonial entertainment inevitably downplays what Schiappa (1999) understands to be the cultural and political significance of the genre. This limiting conception of epideixis as displays of praise or blame proved influential to subsequent rhetorical theorists for centuries (Chase 1961, 297–98).

However, in the twentieth century, rhetoricians began to emphasize the practical and conceptual importance of epideixis (Timmerman 1996). In particular, Perelman and Olbrechts-Tyteca ([1958] 1969), though agreeing with Aristotle's articulation of epideixis as praise and blame, assert nevertheless that epideictic discourse is significant to argumentation because of its ability to strengthen an audience's "disposition toward action by increasing adherence" to certain values and beliefs (50). By defending "traditional and accepted values," the speaker inevitably serves an educational function, persuading the audience by appeals to allegedly undeniable values (51). Similarly, McKeon (1987) returns to the kinship between apodeixis and epideixis to identify what he terms "demonstrative rhetoric," a form of discourse that produces both actions and words (20). Seemingly taking cues from Burke's ([1950] 1969) idea of rhetoric as inducement to action, McKeon (1987) understands demonstrative rhetoric as rousing others to "accept common opinion," to form groups around this opinion, and to engage in actions based on it (20). The reach of this form of rhetoric extends from the arts and sciences to cultural and social institutions (20). In the sciences, demonstrative rhetoric operates as an activity that reveals or produces data because any "assertion or demonstration" (39) can become evidence that influences our "processes of judgment, disposition, and systemization" (20). Thus, distinguishing among demonstrations, justifications, and acts of verification depends considerably on perspective (39). However, by claiming the scope of demonstrative rhetoric to be largely "epideictic [and] not apodeictic," McKeon inevitably reemphasizes Aristotle's distinction of apodeictic proofs from epideictic demonstrations (20).

Despite McKeon's Aristotelian impulse, Prelli (2006) finds productive implications for the future of rhetorical theory in McKeon's and others' reassessment of the distinction between apodeixis and epideixis. This possible common ground, Prelli contends, allows us to contest the sometimes easy distinction between displays of evidence and displays of values by interrogating how all forms of demonstration ("showing forth" or "making known") are "rhetorically manifested displays" (8). He uses the phrase "rhetorics of display" to describe both the rhetorical operations of demonstrations and the demonstrative operations of rhetoric (1). Rhetorical displays are discourses, images, texts, and objects that exhibit "the opinions, facts, and values that manifest the grounds of argument and proof" (8). While a mathematical or scientific proof might not operate rhetorically in the same way as a demonstration of institutional values, both forms of display do operate by way of what Prelli calls "rhetorical selectivity" (12). That is, displays of evidence in scientific contexts and demonstrations of values in nonscientific ones make visible some ideas and attitudes while concealing others (11–12). Any display, no matter the context, is "a rhetorical performance"—"a spectacle to be seen"—presented for the benefit of some audience (14).

While I agree that we should not conflate apodeixis with epideixis by assuming all displays of evidence function like displays of belief, I demonstrate in the chapters ahead that rhetoric and TPC can benefit from a more complete understanding of how these two forms of demonstration make possible the work of science. As I have discussed, scientific, medical, and technical work depends on multimodal displays that operate as data or evidence deployed to support, refute, and communicate facts and arguments. Science also depends on the various epideictic formations that Sullivan (1991) terms "rhetoric[s] of orthodoxies." Sullivan contends that "internal scientific discourse" serves an epideictic function in that many of the inner workings of a scientific or technical field consist of practices that build community by "establishing and maintaining beliefs, values, and ways of seeing," all of which are epideictic operations (232). In a move reminiscent of Kuhn's (1962) description of science as necessitating

epideictic operations in science

tradition and consensus, Sullivan (1991) contends that a particular epideictic formation, a rhetoric of orthodoxies, permeates five aspects of scientific practice: (1) education and training, (2) struggles over legitimation, (3) demonstrations of proof, (4) celebrations of accomplishments, and (5) internal criticism of scientific findings and practices (232–33). In each of these occasions, members of a scientific, medical, or technical community inevitably engage in debating, affirming, rejecting, or communicating attitudes, orthodoxies, and ways of seeing prized by their community.

In medicine's educational and clinical contexts, epideictic performances instill in participants (primarily health-care professionals, but sometimes patients) certain values and ways of seeing. For example, many medical schools conduct public ceremonies for instructors and administrators to bestow first-year students with the symbolic white coats of their profession. (Indeed, they call them white-coat ceremonies.) At these quite formal occasions, entering students stand before an audience to receive this symbolic vestment and recite the Hippocratic Oath, a pledge Keränen (2001) has identified as an epideictic gesture meant to express and support certain ideals (56). Yet epideictic performances of medicine also occur outside of professional settings. Segal (2005) has found that specific medical narratives, such as autobiographical accounts of living with illness, have become an epideictic genre that promotes particular accounts and experiences of illness as more authentic or valuable than others.

While my analysis considers all five of Sullivan's forms of epideixis, *Rhetoric in the Flesh* concerns itself primarily with epideictic occasions found in pedagogical practices, demonstrations of evidence, and ethos-building legitimation strategies. I attend to each of these to illuminate how the epideictic impulse authorizes and structures the practices of this educational space. In fact, I maintain that these epideictic formations of science are most visible in situations of education and training in part because education is a type of prolonged epideictic encounter. Expanding on Perelman and Olbrechts-Tyteca's ([1958] 1969) observation, Sullivan (1993a) conceives of education, generally speaking, as a type of epideixis that employs praise and blame to teach students the reasoning and attitudes deemed "appropriate to professional and public practices" (71). By offering examples of and practice in the values and behaviors to emulate and avoid, the instructor becomes "an orthodox representative" of that discipline who provides guidance and opportunities to engage in the habituated activities of that group (82). Education—specifically, its epideictic impulse—produces knowledge and ways of knowing by socializing students to the practices, discourses, and objects of that group.

My Argument

When I say that multimodal objects and discourses of the gross lab are rhetorical displays, I am referring to the way images and words function as demonstrations of proof or evidence, revealing anatomical knowledge, and as demonstrations of cultural or community values, revealing institutional ways of seeing. These rhetorical processes are particularly important in scientific, medical, and technical contexts, and they are easily visible in the contemporary gross anatomy lab—an assemblage of discourses, displays, objects, and embodied practices that over time develop in participants a trained vision that shapes how they view and respond to the human body. In particular, the cadaveric lab depends on two types of rhetorical displays. The first consists of primarily, though not exclusively, apodeictic displays of visual and multimodal objects that represent and present anatomical knowledge and bodies. The second consists of two rhetorical discourses, two overlapping epideictic formations that in particular authorize and structure those practices: one that praises cadaver-based anatomy as paramount to learning, and one that eulogizes anatomical donors as gift givers.

Of course, apodeixis and epideixis do not always correlate neatly to visual displays and verbal discourses. In anatomy labs, visually oriented multimodal displays primarily serve an apodeictic function; participants deploy them as demonstrations of proof that represent, present, and perform medical knowledge. At the same time, the lab's verbal demonstrations, written documents (like textbooks), and conversations may serve an apodeictic function when participants use them to demonstrate anatomical evidence. Epideixis, on the other hand, is most visible in the overtly rhetorical discourses and stories that participants (instructors and students alike) engage in to convey institutional values and the *doxa* of medicine. I focus on two major epideictic discourses that participants take up to make sense of their laboratory experiences. These epideictic figures, praising dissection and eulogizing cadavers, inevitably structure participants' perceptions of anatomical practices and objects.

The complex multimodal displays, along with other forms of rhetorical demonstration, function as displays of evidence for confirmation-based practices of learning. To learn and teach anatomy, participants engage bodily with the multimodal displays and physical objects that constitute the lab space. Using these objects to view, touch, cut, and understand bodies, participants acquire a trained vision that makes possible not only medical knowledge but also the anatomical body, a conceptual body formed from anatomical discourse and human flesh, one that precedes the patient-body of clinical medicine. The embodied practices of demonstration, dissection, and observation—the primary activities of the lab—enact this anatomical body. The anatomy lab, then, consists of two bodies: the lived body of human experience and the anatomical body of medical science. These two bodies, the physical human body and the foundational biomedical body, form one another. After all, participants learn the second body by projecting anatomical discourses onto the first. The anatomical body, then, is simultaneously subject and object, experiential medium and inscribed surface, biography and personhood. As a result of these habitual, skillful practices performed on and by way of the body, participants develop anatomical vision, a skilled perspective that makes the anatomical body recognizable.

The anatomical body and the anatomical vision students develop to make and make sense of this body are not only the result of embodied practices; the institutional discourses that shape those practices play a powerful role in forming the anatomical body and anatomical vision. From the moment students enter a cadaver lab, the administrators, instructors, and TAs introduce them to an epideictic discourse that (1) praises firsthand observation, particularly cadaveric dissection, as key to anatomical learning; and (2) defines anatomical donation as a worthy, even altruistic act. The encomia of the cadaver, which extols its superiority as an educational object, and the analogy of the gift, which configures the cadaver as a gift and the donor as a gift giver, operate as rhetorical frames that participants use to filter their experiences with cadavers. These two epideictic discourses inevitably reinforce the values and ways of seeing of anatomical science and medical education.

Through the apodeictic displays and the epideictic discourses, participants acquire clinical detachment—or what Fox (1979) calls "detached concern" (56). Yet rather than see detached concern as depersonalization, I contend we should view this phenomenon as a sensitization, a trained perspective that puts the science and the personhood of the body in constant and productive tension. This process of attending sensitizes participants to the simultaneous biological value and humanity of the body. I offer this concept as a modification of traditional clinical detachment, one that recasts the cadaveric body (and the anatomical body) as a contradictory object that never completely loses its humanity. Through their interactions with anatomical knowledge, multimodal displays, and embodied practices, as well as the institutional epideixis that supports the lab's activities, participants come to understand cadavers and all anatomical bodies as medical and personal.

Rhetoric in the Flesh, then, demonstrates how bodies, multimodal objects, and discourses coalesce to form an embodied rhetorical action, an object-body-environment assemblage that exerts a socializing force that structures human experience. Embodied rhetorical action is the habituated use of objects, images, and discourses in a distinct social field that allows participants to deploy them effectively. Deploying these objects and discourses makes possible perceptual attitudes and dispositional tendencies that provide the structure by which these activities are authorized and made meaningful, allowing them to influence future actions and attitudes.

The Book in Outline

According to an anatomy professor of my study, the body is a challenging space to enter: "There is no place where the body begins. It is a circle. It is difficult to find a way, a suitable place to begin dissection" (William, instructor of both courses). Not simply a text or a transparent object, not exactly a window or a doorway, the body is more like an enclosed loop, a circle, a process. In large part, I have organized the book to bring this formation of the body into greater focus. Like the act of dissection, I begin with visible surface structures, then slowly proceed inward to the more delicate interior ones. Like the human body, where structures lie one atop the other, the lab's activities, both the embodied practices and the rhetorical ones, overlap in their interconnectedness. These activities are so dependent on each other that they often happen simultaneously. To capture these coconstructed actions, I have, for better or worse, constructed *Rhetoric in the Flesh* as a book-length argument, while still seeking to give each chapter the integrity necessary to stand alone. This decision means that some chapters closely link to and even anticipate others. In chapters 3 through 5, I explore the activities that structure learning—looking, touching, and cutting—because these are the skills students and TA must master. The later chapters turn to the epideictic discourses and formations that support and structure those practices. I do not mean to imply through this organization that apodeixis and epideixis are easily separable. My hope is that when I explore epideictic formations in chapters 6 and 7, their connection to the apodeictic displays will be even more apparent.

In chapter 2, I describe in detail the anatomy laboratory classes that were the site of my fieldwork: the undergraduate lab, the medical and dental student lab, and the preparation labs where the teaching assistants set up each week's materials. In the process, I introduce the phenomenological and cognitive components of my theoretical framework. Via this framework, I demonstrate how bodies and multimodal objects form embodied rhetorical action, a body-world coupling that exerts a socializing force that structures how participants experience and understand these objects. Throughout the book, I use and continue to elaborate my theory of embodied rhetorical action in order to explain how discourse, objects, documents, and bodies converge to make possible anatomical vision and anatomical bodies.

In chapter 3, I attend to the practices that constitute anatomical presentation and representation, or demonstration, by examining the lab's networks of multimodal objects that bring the anatomy body before the eyes and hands of laboratory participants. By classifying the multimodal displays according to their representational content, dimensionality, materiality, and perceived level of interactivity, I explain how these displays and the embodied rhetorical practices they encourage ultimately configure a trained perspective based on the affordances of the objects on display. Students and TAs conceive of these representational and presentational objects as enacting both anatomical knowledge and opportunities to act on and with the body. Building from this discussion of affordances, chapter 4 takes up how participants learn anatomy through the process of observation, especially the capacity to recognize anatomical structures in and on the physical body. Students, TAs, and instructors use the lab's displays and objects to learn and teach anatomy through the process of hypothesis confirmation, a process that entails learning to recognize the descriptive and relational evidence of the anatomical body. By developing awareness of how a structure should look (its descriptive aspect) and its associations with neighboring structures and basic physiology (its relational aspect), students learn to recognize the body's anatomy by learning to recognize and be persuaded by anatomical evidence.

Dissection is a process of revealing to learn and learning to reveal. Through it, participants make the anatomical body visible, all the while causing the physical body to slowly disappear; they adopt an almost aesthetic orientation that allows them to view the well-dissected cadaver as a "beautiful body," made attractive by its ability to become an authoritative presentation of actual anatomy. In chapter 5, I illustrate this by way of field notes and interviews as well as an analysis of Gunter von Hagens's *Body Worlds* exhibit of plastinated human cadavers, a display that (for most participants of my study) epitomized this seemingly contradictory aesthetic category. Expert dissection not only renders the unruly cadaver (in the words of participants) "beautiful" and "clean," but it does so by encouraging dissectors to use the idealized atlas images as an interpretative framework for understanding the body and organizing the work of dissection.

The lab's two rhetorical demonstrations—the apodeictic multimodal displays and objects as well as the epideictic displays of institutional values—develop in participants what is commonly termed "clinical detachment." This mechanism allows them to filter (as much as possible) the unpleasantness of dissection and the anxiety producing nature of cadaveric anatomy. Chapter 6 calls into question the classic idea of clinical detachment as merely a desensitizing influence. Through the process of anatomical focus (the transforming of the physical body into the anatomical one), participants learn to navigate and even sustain the tension between the body as science (biology) and the body as person (personhood). The powerful institutional rhetoric that praises cadaveric anatomy, specifically firsthand dissection, makes anatomical focus possible. This encomium of dissection structures participants' experience of working with cadavers by encouraging them to hyperfocus on the actions of dissection (as opposed to the object of those actions) and the scientific usefulness of the body. This praise of cadaveric dissection is not the only epideictic discourse to structure the experiences and practices of the lab. In fact, the entire process and philosophy of anatomical donation is a result of an epideictic analogy of the gift. This discourse of benevolence, as I discuss in chapter 7, constitutes the body as a valuable gift to scientific knowledge, medical education, and the future of health care—a gift with biomedical value. Participants inevitably filter their experience with cadavers through this body-as-gift and donor-as-gift-giver formulation, allowing them to make peace with dissecting—and thus destroying—the human body. This analogy of the gift influences participants' daily interactions with their assigned cadavers as well as participants' thoughts about self-donation.

The convergence of visual displays, embodied practices, and rhetorical discourses together form embodied rhetorical action, a type of intercorporeal connection with the world around us. Through the ontological, rhetorical, and socializing force of embodied rhetorical action, participants of the gross lab come to see the physical, material body as the anatomical body of medical knowledge. In the book's conclusion, I elucidate the implications of my work for other TPC contexts. Because I am addressing audiences in TPC and rhetorical studies primarily, I end each chapter with a brief discussion of my work's implications for our research, teaching, and practice. Ultimately, my work aims to elucidate the mutually constitutive relationships among expertise, trained vision, and rhetorical displays. To understand how a participant learns a skill, we must consider the embodied practices involved in becoming an expert. But a focus on embodiment will not provide the entire picture. To understand how someone becomes socialized to see objects and people by way of the frameworks provided by a profession, we also must take into account the rhetorical discourses that structure those objects, displays, documents, and practices.

2

ONE BODY TO LEARN ANOTHER

Activities of the Anatomy Lab

Even before entering their first anatomy lab, students no doubt conjure up images of what they might find there. We, of course, would do the same. We might envision, first and foremost, cadavers, tables of bodies set out for others to dissect and explore. We might think of bright laboratory lighting, black countertops, shiny metal tables, and posters of complex biological systems. We might consider the smell of formaldehyde emanating from the dead man or woman lying nude beneath some spotlight of fluorescence. We may picture the training of doctors and dentists who spend what must be years in those cold labs, performing autopsies on the dead and cutting, layer by layer, through the entire body. At least, these are images that horror films and police television shows have offered us for decades. But if we were to visualize the labs in this way, we would only be partially correct. There are cadavers. Preserved, often in a solution of formaldehyde, phenol, and isopropanol in water, they are wrapped in either muslin or plastic or both until they need to be revealed. There are bright lights and gleaming metal, but there are also walls decorated with colored posters and whiteboards filled with diagrams and reminders. And the rooms that wait on the other side of a punch-code lock are not particularly cold, do not look like a crime lab from the television show *Crime Scene Investigation* (*CSI*), and most definitely do not exhibit the fetid squalor of Dr. Frankenstein's laboratory.

The contemporary anatomy lab (figure 2.1), more a classroom than a laboratory, is the space where future doctors, dentists, physical therapists, and nurses encounter the cadaveric body of traditional Western medicine—the body that is anatomical knowledge made flesh. Although not searching for the cause of death, students do cut into the body in order to see its interior, to recognize and learn an array of human structures by slowly and methodically anatomizing that body to nothingness. They dissect the body until nothing is left.

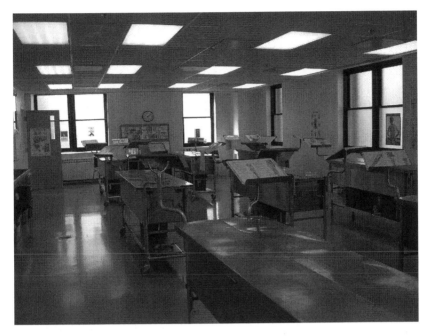

FIGURE 2.1 Anatomy laboratory space with cadavers concealed.

What we would no doubt fail to realize is the role the lived human body plays in the gross anatomy lab. We might also overlook the formative roles that rhetorical processes of demonstration and persuasion play in this educational setting of burgeoning expertise. In *Rhetoric in the Flesh*, I explain how the work of the anatomy lab and the trained vision at the center of that work is constituted by the rhetorical actions of lived human bodies. Through demonstration, observation, and dissection, activities at once embodied and rhetorical, participants develop a powerful trained perspective that shapes their technical expertise and their conception of the human body and that also influences their interactions with other bodies.

Two gross anatomy laboratory courses were the site of my fieldwork: (1) an undergraduate prosection course organized by the viewing of already dissected cadavers, and (2) a more advanced dissection course for first-year medical and dental students. In this chapter, I discuss these labs as an assemblage of discourses and practices, of images and objects, of eyes that look and hands that touch. In these spaces, students and teachers use the physical body to learn the anatomical one. And to understand this work, we must understand the lab's discursive and material practices as well as the ways these practices are involved in a form of embodied rhetorical action that is visually and haptically directed and that is central to the socialization of participants. By sketching out the typical lab

related to the sense of touch

practices I studied, in this chapter I provide an overview of the surface anatomy, if you will, of anatomical vision. In subsequent chapters, I excavate and explore in detail the intricacy of these practices that enact the vision and expertise of anatomical education.

Anatomical Bodies—Objects and Subjects

> Since you will be keeping company with your body for another 60 or more years, you had better know more about it than this! Anatomy is a fascinating and relevant subject for study. The more you involve yourself, the more you'll get out of it. ("Student Survival Guide" to prosection course)

This statement, from a handout that accompanies the syllabus in the prosection course, posits anatomical knowledge as educationally and personally relevant because students will live in their bodies for the rest of their lives and, thus, should know more about the bodies that they are. Notice that anatomy is not praised as merely a means to a professional end; instead it is "fascinating and relevant," whether or not students ever become medical professionals. While most students in this course, as in any anatomy course in the United States, are seeking a career in some medical profession, this introductory course positions itself as valuable because it offers knowledge that is useful to the embodied humans taking the course. The last line puns on students' intellectual involvement and their physical bodily involvement, both of which are required to excel in the gross lab. Students engage their physical bodies to learn the anatomical one, a conceptual body manifested in cadavers, displays, and their own living flesh.

From the beginning, the undergraduate prosection course establishes itself as involving what Varela, Thompson, and Rosch (1991) describe as "the two sides of embodiment," namely the body as object and the body as subject, or having a body and being a body (xv). These notions of the body, which structure participants' experiences of the lab, have consumed philosophers since Descartes ([1641] 2000) supposedly split the mind from the body in *Meditations on First Philosophy*. In particular, the branch of philosophy known as phenomenology has been motivated to explore what it means for the human body to be both, as Husserl describes it, *Körper* and *Leib*, or a material (even dead) object and a living subject (Leder 1992, 24–25). Merleau-Ponty (1968) refers to these "two sides of our body" as the "objective body" and the "phenomenal body." The first is our body as a "sensible" object that we (and others) see, touch, and experience as a thing in the world; the second is our body as a "sentient" being, or subject, by which we experience the world around us (136). Merleau-Ponty ([1945] 2005) contends that we are objects and subjects simultaneously (239); and the "intertwining" of these two forms of embodiment are the basis for our perceptions and, thus, our capacities for thought and judgment (Merleau-Ponty 1968, 138).

The intertwining of these modes of embodiment structure the practices of the gross lab, as students, TAs, and instructors engage their entire bodies to recognize, confirm, debate, teach, and communicate anatomical knowledge. Participants use their bodies to learn the anatomical body, and they do this by learning anatomical knowledge and reading that knowledge in the bodies of the lab—the living, the dead, and the multimodal displays used to represent them. Anatomical knowledge becomes the lens through which they view the body as they engage in anatomical practices of demonstration, observation, and dissection. In the gross lab, bodies are objects and subjects, products and processes, involved in a form of embodiment that facilitates the gross lab's discourses and practices.

Discourse and Practice

Anatomical knowledge is a set of concepts, terms, and orientations for describing the structure–function relationships of biological bodies. The TAs and instructors I studied understood one of their objectives to be helping students keep up with, make sense of, and learn what they often described as a new language. Although anatomy lacks the linguistic grammar of a formal language, discussing and understanding anatomical discourse as a language of the body is quite common. Anatomically speaking, the term "leg" designates the area below the knee, whereas the "thigh" is the area above the knee. One easy, perhaps tricky, question on the first exam might ask students to identify the upper region of the leg (above the knee), using only the anatomically correct answer (the thigh). Though a trivial example, this renaming of a commonly referenced body part illustrates the way anatomical terminology often reinscribes the colloquial expressions for our body. Anatomical discourse inscribes the body with names as well as movements and directions, in part to aid medical professionals in specifying more precisely the location of structures. The cadavers and the various multimodal displays of the anatomical body are positioned and described according to the anatomical planes (coronal, sagittal, and transverse). In figure 2.2, a photograph of a poster from the undergraduate course illustrates one popular rendition of what is called the anatomical position. This adaption of Leonardo da Vinci's *Vitruvian Man* displays anatomical planes while implicitly connecting the contemporary lab to anatomy's rich historical tradition, in this case the anatomical innovations of the European Renaissance.

This discourse of anatomical terms has real material effects both in renaming the body and in authorizing certain ways of knowing that body. No healthcare worker could function in a Western medical setting without this knowledge, which structures how practitioners view and work with patients. Young (1997) argues that medicine inscribes the body with anatomical concepts in a process that subjects that body to the domain of science, thus translating bodies from the "realm of the ordinary" to the "realm of medicine" (46). As a representational

[handwritten note:] → authorizing certain ways of knowing the body / ways of knowing.

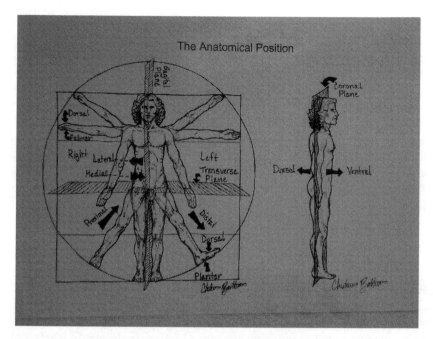

FIGURE 2.2 Illustration of the anatomical position and major anatomical planes.

system that provides (supposedly) authoritative knowledge, anatomy exerts what Foucault ([1978] 1990) describes as a disciplinary power, part of the "anatomo-politics" of biopower that seeks to control and optimize the body's capacities (139). In the lab, the cadaveric body—lying either supine or prone, often with sections completely cut away to expose structures beneath—becomes an object of anatomical science as lab participants inscribe those bodies with anatomical knowledge and thus view them as instantiations of that knowledge made flesh.

Anatomy is also a set of practices that make the anatomical body possible by making it visible. Participants map anatomical language onto dead and living human bodies through the activities of demonstration, observation, and dissection. In the process, students learn the skills and embodied know-how necessary to properly view, touch, and cut cadavers, transforming these bodies to reveal or materialize anatomical knowledge. In this sense, participants make the anatomical body through the interplay of discourse (knowledge) and physical practice (know-how), which prefigures much of their future medical careers, specifically the enactment of medical bodies through discursive and material means. In medical contexts, whether the hospital or the gross lab, we have a body, and we are a body. And through the discourses and practices of medicine, we "do" our bodies; we "enact them" (Mol and Law 2004). As Mol (2005) shows in her ethnography of clinical practices involved in treating atherosclerosis, objects, specifically bodies, and the concepts necessary to make sense of them emerge through practice.

The practices we exert on objects structure our conception of and relationship to those objects. That enaction occurs, I contend, through embodied rhetorical activities, supported by epideictic and apodeictic formations.

Epideixis and Apodeixis

From the beginning of the course, instructors introduce students to the bodies and discourses that will structure their actions and eventually shape their perceptions. Besides anatomical discourse, students also experience subtler rhetoric discourses that instill in them the cultural values and orthodoxy of the gross lab. These formations of epideixis enact the body as simultaneously an object and a subject, both dead material and living flesh. This inculcation and socialization begins with students' first exposure to the cadaveric bodies on the course's first day: "There are 50 cadavers for the course. That means 50 people donated their bodies," says the instructor of the undergraduate course to the students clustered around the table-like cadaver tanks in the first laboratory room. "We require you to show respect for the human material," he adds. "These people donated to teach anatomy, and not to be a spectacle for someone off the street." He admonishes them not to invite unwanted guests to the lab. "No friends, no family, no relatives," he repeats. "No uninvited guests." With this warning, the instructor introduces the cadavers as valuable resources for anatomy education and invaluable donations offered by once-living people. The cadaver must be respected as an object crucial to their training, a gift for the purposes of that education. This dual articulation of the cadaveric body as a specimen and gift runs throughout both anatomy courses I observed, as well as the larger discipline of anatomical science. As I explain in later chapters, these epideictic formations facilitate a unique form of clinical detachment that many participants experience in the gross lab.

The bodies of the lab, however, are not merely dead objects of study or honor. Bodies crucial to the lab are also the living ones of participants; they use the skilled capacities of lived experience to enact the anatomical body. As I demonstrate throughout this book, students' interactions with cadavers encourage them to appreciate forms of embodied knowledge required for learning through situated practice. For example, on that same first day of the undergraduate course, the director of the anatomical bequest program addresses the emotional component of working so closely with the dead. "If you are feeling anxious," he tells them, "it is normal." He acknowledges that for many of them this will be their first exposure to "the cadaveric body." "If you are feeling uncomfortable, please email me or call," he offers. He also encourages them to talk to the course instructor. "And if you feel lightheaded, you should listen to that." He continues, "What I tell people is to listen to their body." He reminds them to eat breakfast and to give their body some sugar. "But it is O.K. to get a bit flushed or to need to have a seat." It is "understandable," he says, if they need to take a break in the hallway sometimes.

Here, we see an authority figure giving students' permission to feel the more troubling or, at least, conflicting experiences involved in cadaveric anatomy; we see him advising students to "listen to their body," to detect and respect the subtle and not so subtle clues their own body provides. Though he is referring specifically to physical and emotional feelings of unease (including perhaps dizziness and nausea), the bequest director's words highlight the role of embodied knowledge. Specifically, in order to make sense of the sometimes bewildering multimodal displays that surround them, students must learn to recognize and evaluate the anatomical evidence they gain from the bodily experiences of seeing, touching, and sometimes cutting cadavers. Through practices of viewing, touching, and cutting, participants develop a rhetorical and physical ability to read the various multimodal displays as examples of anatomical knowledge. The plastic models, x-rays, plastinated bodies, and atlas images (or plates) display anatomical knowledge to those participants intellectually, physically, and rhetorically trained to access and enact this knowledge. These multimodal displays and objects do not merely represent the anatomical body; they enact that conceptual and authoritative body that for centuries has delimited, demarcated, and inscribed the physical human body. But to enact this anatomical body, students must learn how to physically interact with the lab's multimodal displays—how to properly see, touch, move, and read them—in order to recognize and verify the evidence these apodeictic objects exhibit.

Cadaver-based anatomy education inscribes the human body as an object of study and conceives of the body as a vehicle in which to study that object. Thus, participants are taught to read the physical body as text, to learn the anatomical body as subject matter, and to do so by using their own bodies as instruments. In the gross lab, anatomical knowledge is discourse made flesh or enacted in the flesh through practices that are embodied and rhetorical simultaneously. In order to make this point clear, I provide an overview of the material activities of the spaces I studied: (1) a prosection-based undergraduate lab, (2) a medical and dental dissection-based lab, and (3) the preparation labs where TAs set up each week's materials.

The Undergraduate Course: A Prosection Lab

Anatomy 302 (or ANAT 302) is a lab course consisting of demonstrations and observations of prosected human cadavers, meaning students in the undergraduate course work with previously dissected bodies. Thus, students in the course do not dissect. Instead, they look, touch, handle, and verbally explain the cadavers in order to fulfill the major course objective, which is (according to the syllabus) "to provide an introduction to human anatomy with emphasis on relationships of structure to function." To accomplish this objective, instructors and TAs require students to take the lab concurrently with a separate, more conventional lecture course (ANAT 301) that covers basic physiology. Yet, more than a supplement to the lecture, the lab is a space of embodied learning quite different from the lecture in that students use their bodies to learn the body. The actions of course participants are organized by three major activities: (1) prelab talks, (2) in-lab viewings, and (3) study time.

Prelab Talks

Each session of the prosection course begins with a prelab lecture, or an informal presentation by an advanced TA or instructor who provides an overview of that day's material. This presenter stands at the whiteboard and presents the detailed list of structures on display in the next two rooms, using that week's identifica-   *Taxonomic systems* tion list, or ID list. These lists, which correspond to the content of the ANAT 301 lecture taught by another instructor, arrange the body by anatomical systems (for example, respiratory, digestive, reproductive). The act of naming structures and briefly discussing either their function or providing a helpful mnemonic sets the agenda for these brief informational sessions. The prelab talk is the only group presentation students receive in the prosection lab, and it is structured less like a lecture and more like a forecast, previewing structures students need to view and learn. Even before the cadaveric demonstrations, anatomy education begins with the presentation of the boards, the course packet, the ID lists, and the presenters' bodies, as TAs and instructors often direct the students' attention to their own surface anatomy. For example, during the prep-lab talk on the muscular system, a male TA propped his leg on the table in front of him, pointing out the muscles   *✳* of his calf. This brief presentation of the TA's own body visually oriented the stu-   *body as* dents to the physical human body as the site of anatomical knowledge, reminding   *an* them to attend to the body—whether living or dead—as an anatomical object.   *anatomic-* After the prelab lecture, students move into the spaces set up with cadavers.   *al object.*

Lab Viewings

Most of the teaching occurs during the demonstration periods. Again, TAs teach students to tune in to features of the human body as illustrations of anatomical knowledge. Typically the students move from table to table, listening to TAs present structures. During these presentations, a TA might point to a structure on a cadaver (or an excised bone or organ), name it, and offer a mnemonic or a description of some relationship or neighboring landmark. TAs provide this information so students can recognize the structure, remember its name, and identify it later. These lab demonstrations are not merely receptive acts of passively seeing and recognizing, which are themselves, of course, physically, socially, and cognitively active endeavors; instead, students must be prepared to take part in and lead these demonstrations when asked.

This portion of lab is loosely structured around the independent movement of students (individually, in pairs, or in small groups) through the lab rooms, allowing them to spend more time at some stations and less at others. TAs often stand in the middle of four tables positioned to create a circle around them. These TA corrals, as they are called, allow students a better vantage point on the displays and cadaveric prosections on the tables as well as the TA who moves inside the corral, giving each student an optimal view of the structure the TA is teaching (figure 2.3). In a sense, the corral not only allows students a better view, it also grants the TA a greater range of motion to reach objects on the tables.

On an ordinary day, a group of students stand no more than two rows deep around a particular corral, listening to and watching a TA's demonstration. The students' eyes move back and forth between the TA's eyes, the multimodal object on display, the ID lists, and their notes.

During these lab demonstration sessions, cadavers rest atop the closed tanks, offering what one TA, Nicole, described as the purpose of the lab—"to see the actual structures of the body, instead of just textbook learning." As 3-D multimodal displays, cadaveric specimens allow participants to investigate the same anatomical structures on different bodies. TAs and instructors often tag each cadaver with labeled string that marks muscles, arteries, and veins so participants can use each cadaver to demonstrate several features. By tagging muscles, for instance, across a number of cadavers, TAs call attention to anatomical variation. TAs and instructors display both tagged and untagged cadavers in a number of

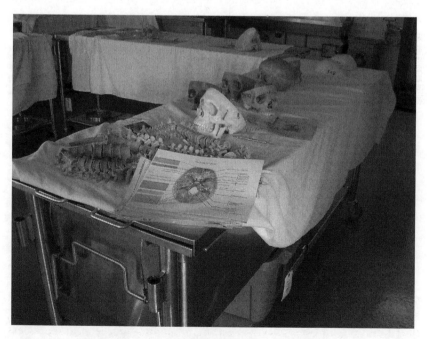

FIGURE 2.3 TA corral with bones and illustrations.

ways and in various stages of dissection in part because cadavers offer illustrations of particular structures and not others. For instance, a male cadaver might lie on his back with his head and face exposed, and the head of the female cadaver beside him might be wrapped in muslin or cloth. Another male cadaver close by, with tattoos visible on his torso, might be missing his lower limbs. No one body can clearly present every anatomical structure because the human body is a complex object (and subject, for that matter) comprising layers of intertwining structures. Also, the bodies of the undergraduate prosection course are "second use" cadavers, meaning they have been partially dissected by students in another course. In this case, a physical therapy program previously used them to learn primarily muscles before sending them to this prosection lab.

Because of this presentational variation (from cadaver to cadaver and across types of multimodal displays), instructors and TAs encourage students to ask clarification questions about the location and function of structures and their possible mnemonics. At the beginning of the semester, students usually begin these question-and-answer sessions by asking whether or not something is true. A student might point to a cadaver and ask, "Is it true that when you sprain your ankle, this structure here is what hurts?" These is-it-true questions attempt to connect anatomical structures to ailments or diseases and allow students to play the role of doctors, thus mimicking the practice of diagnosis that will become an important component of their future work as medical professionals. These questions also inevitably seek out the clinical relevance of the cadaveric body. By viewing cadavers as a possible example of disease or a resource for making sense of an ailment, students substitute the cadaveric body for the patient-body. In these moments, the cadaver becomes not just the anatomical body—the instantiation of anatomical knowledge—but also the stand-in for the patient. Interestingly, TAs frequently respond to these questions by reminding students to focus on the anatomy first, to use the lab to understand basic anatomy before they concern themselves with diagnosis. In other words, the anatomical body must be recognizably visible in the cadaver before the patient-body can be viewed. Throughout the semester, students learn to attend to the cadaveric bodies, and thus the anatomical body, by becoming attuned to the types of inquiry appropriate for learning anatomical knowledge.

Study Time

> wow - weren't those bodies?

As students examine these multimodal displays, they take notes in their course manuals, discuss what they see, and quiz each other on structures, functions, and mnemonics. Quizzing and being quizzed by TAs or peers is an exercise that dominates the last portion of the lab. After the TAs finish presenting structures, students walk around, usually in pairs, examining cadavers and displays on their own. During a class session, I watched a TA, Jerome, ask a group of students many identification questions. "Who knows what this line is called?" he asked as he pointed at the cadaver in front of him. The students physically leaned closer to the body for a better look. "If you learn the bone sections very well, the muscles come naturally," he told them. A student tentatively used what he remembered to be the mnemonic for the psoas muscle: "Something about psoas for your sore ass." Jerome replied with a positive acknowledgment: "Yes, that is the psoas." Here a TA quizzes students to help them learn by gauging how well they understand the material then by asking questions that match that level of knowledge. If the students answer the questions correctly, the TA moves on to a more difficult question. To do this, a TA must attend to the students' level of comprehension and make explicit to them what they do and do not know, thus offering students a chance to observe and assess themselves and each other. During individualized study times, students quiz each other and demonstrate again the structures as TAs have done.

The Medical and Dental Course: A Dissection Lab

The medical and dental course is a more advanced version of the undergraduate one. In it, students learn anatomical structures to understand their clinical relevance. According to the syllabus, instructors require the medical and dental students do the following:

1. Know "the major anatomical structures of the human body and their primary functions."
2. "Recognize important clinical structures and landmarks on various radiographic images including CT, MR (magnetic resonance) and x-ray images."
3. "Recognize and/or palpate, based on an understanding of surface anatomy, various anatomical structures, both superficial and deep."
4. "Make reasonable predictions of the clinical manifestation of injury or disease to anatomic structures."
5. "Relate anatomic structures to clinical diagnostic procedures and treatment approaches."

Because instructors, and arguably the curriculum designers of the medical and dental schools, understand this course as foundational to their professional education, students learn to view bodies and at times touch them (notice the reference to palpation) as a clinician would, namely for the purposes of diagnosis. Through dissection, demonstration, and observation, medical and dental students learn to attend to the clinical relevance of the anatomical body by attending, through vision and touch, to the physicality of the body and the opportunities for learning it provides. It is no coincidence that the first three goals stress the importance of recognizing and knowing anatomical structures before seeking out their clinical and diagnostic relevance. Again, the anatomical body is the first medical body exhibited in this space.

As a training ground for future physicians and dentists, the dissection lab offers students and TAs the opportunity to become surgeons-in-training by training not just their minds but their bodies. The experience gained through vision and touch, the development of manual dexterity, the tuning of attention to recognize relevant anatomical structures in and on the human body, and the embodied experience of simultaneously working on and inhabiting a body: these factors contribute to learning in the dissection lab. Much of this learning is communicated through self-study skills students acquire through performing dissections in "body-buddy" teams of four. Dissection allows students to explore cadavers on their own, offering an awareness of the body's three-dimensionality as a system of relationships layered one atop the other. Because clinical relevance is crucial, students must painstakingly dissect the body to understand the interconnectedness of parts. Students in this course are evaluated by way of three written exams and three practical laboratory exams, the latter of which are similar to the undergraduate exams. The major activities of the course are (1) lectures, (2) prelab talks, (3) dissections, (4) prosection presentations, and (5) study time.

[margin handwriting: How to use the body. Manual dexterity etc. …]

Lectures

Unlike the undergraduate prosection lab, this eight-week dissection course comprises formal lectures and dissection sessions. In the morning, students attend a conventional lecture, in which a course instructor presents information on that day's dissections. Whereas the undergraduate lecture course is a separate course that covers anatomy and basic physiology, the primary goal of the ANAT 600 lecture is to prepare students for the large number of structures they must dissect, observe, and demonstrate. Through the use of PowerPoint presentations of atlas images, CT scans, x-rays, photographs, and other illustrations, lecturers preview structures that students should find and basic dissection techniques they will use. As such, these lectures offer an overview of the anatomy knowledge students must learn and the tasks they must perform.

To illustrate the clinical relevance of structures and offer visual and embodied mnemonics, the lecturer often gestures and demonstrates with his or her own body. For example, any lecture on major muscles inevitably involves physical demonstrations to illustrate how a muscle works, calling attention to anatomical structures on the living body. These routine self-demonstrations, or anatomical performances as I call them, are significant because they specify the body as the location of anatomical knowledge while inscribing the lived body as anatomically significant. Instructors conceive of these lectures as review sessions in that they expect students to have previewed that day's material before coming to lecture. To prepare, students find a wealth of online information, including that day's lecture notes and slides, on the courseware instructional site. And though only half of the medical and dental students I interviewed mentioned studying before lecture, nearly all of them (and almost every student in the lecture hall) printed

[margin handwriting: physical demonstrations]

the slides and reviewed them during the presentations. The TAs usually sat in the back, taking written or mental notes, particularly noting (as all mentioned during our interviews) how the lecturer described the structures. So as to avoid giving confusing or contradictory information, TAs attempt whenever possible to reiterate the same language, examples, and mnemonics during their discussions with students.

Prelab Talks

After the (typically) 50-minute lecture, students quickly make their way to the locker rooms (one for men and another for women) to change into scrubs and other "work wear" (such as old T-shirts and sweat pants). Like the undergraduate prosection course, this dissection lab begins with a prelab talk by one of the two TAs assigned to each lab room. These talks focus on practical reminders of dissection techniques, locations of structures, and helpful mnemonics, thus emphasizing important points from the lecture. Several medical and dental students I interviewed understood these five-to-ten minute talks as a condensation of the lecture and a kind of pep talk. Like the undergraduate lab talks, these prep talks consist primarily of what Collins (2010) refers to as "second order rules" (62), rules that provide information about the difficulty of the task at hand, and "coaching rules" (63), rules that provide hints and suggestions for accomplishing those tasks. In the dissection course, the subject of these talks is usually the dissections students will perform. By informing students of both easier and more challenging components of a dissection, or by giving them tips for locating, identifying, and cutting structures, TAs and instructors guide the students' physical actions and teach them anatomical structures. Expressing a widespread conception of these talks, one dental student, Marianna, finds them to be a "nice refresher, a kind of ease in stress" that makes her "feel better" about the material and the work to be done, which can seem "overwhelming" after the detailed lectures.

These brief prelab talks are similar to those in the undergraduate course and inevitably serve instrumental and affective purposes. However, the major difference between the two labs—one a prosection course, the other a dissection course—adds a different dimension to the lab talks. For the prosection course, these talks are agenda-setting meetings and reminder reviews; for the dissection course, they contain both of these elements while also offering motivation. TAs and instructors provide encouragement in the form of personal stories of commiseration. For instance, one TA, Allen, reminded students during a prelab talk to be careful not to destroy nerves. He did so by confiding in them what a difficult time he had preserving nerves when he performed that same dissection as a gross anatomy student. During a talk on the superior back, a TA, Abby, told students not to reflect (or pull back and away) the rhomboid because it is the structure that holds the others in place. She provided a memorable example of watching a group of students do so, to nearly disastrous results, when she took the course. Most of these brief narrative moments offer instruction and lighten the mood while acknowledging the difficulty of dissecting the body. And, as I discuss in chapter 6, they also reinforce cultural values involving the experience of dissection.

Dissection Time

Most of course time is devoted to careful and painstaking dissection and the pedagogical presentation of prosections. During dissection, students work in their independent body-buddy teams, with two TAs and one instructor per room moving about to oversee the work and answer questions. In order to dissect, students must first orient themselves to the position of the body and the location (in the body) of the structure on the ID list. These processes of orientation and positioning involve physical manipulation of the cadavers and themselves in order to visualize the anatomical body before them. The process of visualizing anatomy involves more than studying cadavers or photographs and then calling up a mental image of that object. Instead, students must deftly engage the multimodal displays that represent and present the anatomical body. Participants must ensure that their memory of the cadavers is accurate and sophisticated enough to account for natural anatomical variation. To do this, students learn to materialize the anatomical body through active, physical, tactile explorations of cadaveric bodies and sometimes their own. By moving their hands, eyes, and attention into and across the multimodal displays—atlases, whiteboards, cadavers, themselves—participants physically encounter and explore these displays and, by so doing, comprehend the anatomical body as a 3-D space.

From this understanding, whether limited or vast, students must then transform the cadaveric body in front of them into a presentation and demonstration of the anatomical body. Through a slow, methodical process often described in archaeological terms, students excavate cadaveric flesh to uncover specific anatomical structures on the ID lists. During the dissection period devoted to the forearm (the arm below the elbow), for example, students are required to reveal the muscles responsible for hand movements, such as the flexor digitorum superficialis. But to find this muscle, students must dissect layers of skin, fat, fascia, and other muscles to reach this particular structure in question. By the term "structure in question," I mean the actual structures students are in the process of finding, as opposed to structures in the body they are not searching out. In a kind of figure-ground relationship, the structure in question is surrounded by other structures, some useful as landmarks and others unnecessary items to be removed. When a student looks for the flexor digitorum superficialis, this structure becomes the structure in question; the structures between her gaze and the muscle, those obscuring that view and not on the ID lists, are dissected away. To dissect a cadaver is to use tools and objects to find and recognize structures in

Dissection involves mapping (handwritten marginal note)

question. Because particular structures—such as the urethra, the occipital triangle (of the neck), the heart—are located in the cadaver, participants learn to use tools that allow them to recognize those structures and view them as anatomical and meaningful based on the situated actions that give them their meaning. Context and purpose are key because the structures in question change from lab to lab. In the process, participants render the anatomical body visible by transforming cadavers into 3-D multimodal displays of particular structures in question.

Dissection, then, involves more than cutting into or penetrating the cadaver (both bodily practices); it also involves mapping anatomical terms onto the body so as to read it as anatomy, a process of inscription and incorporation that involves gesture and physical demonstrations and one requiring students to physically and visually orient themselves to the cadaver in front of them.

These practices that enact the anatomical body require students to use several resources simultaneously. To learn, students move physically back and forth between cadavers, atlas plates, whiteboards, and images posted on the walls and doors. This use of multiple resources is necessary to render the body an anatomical specimen, to demonstrate the body as a system of structures waiting to be revealed and learned. Students learn to use their bodies and the lab's resources as embodied tools, which they incorporate into their bodily perceptions through repeated use which, in turn, becomes easier the more they are able to skillfully and seamlessly incorporate these tools into practice. These tools—from forceps to textbooks—become an extension of their body, used to navigate these laboratory spaces.

Prosection Presentations

The second major component of the dissection course, occurring concurrently with dissection, is the prosection demonstrations in which one team uses its cadaver as a teaching tool (as a text and an instrument) for the rest of the class. Before each lab session, TAs carefully dissect a section of a cadaver, creating a more representative prosected model of that day's dissection. The team whose cadaver is prosected must then teach those structures to their classmates. Classmates come over (one team at a time) and listen as the team demonstrates the structures on the ID list. Because the body often contains paired structures—two arms, two legs, two eyes—TAs only prosect one of the pair, for example, one shoulder. Team members take turns teaching the prosected example and dissecting the parallel structure—the other shoulder. But before the prosection demonstrations can begin, TAs review and teach the prosection to the body-buddy team.

During a prosection presentation in this course, I watched a female TA, Marilyn, present on the prosected body, and demonstrate that day's anatomical structures with forceps (which resemble tweezers) to the body-buddy team around her. As she spoke, she looked at the board in front of her and the drawings of the brachial plexus on the easel, calling students' attention to the body and the board. She reminded them of how easy and common it is to confuse the names of veins and arteries. The way to avoid this, she told them, is to say anatomical names aloud to each other. After Marilyn had demonstrated the structures on the ID list, a male student began the demonstration. Marilyn interrupted only to correct and reinforce the names: "Yes, that's it. Axillary nerve." All four students and the TA leaned into the open body to better see the structures. At one point, the male student paused, seemingly confused. "The next one is the subscapular artery; this one here," Marilyn pointed out, holding it with forceps. "Up here is the lateral cord of the brachial plexus, and here is the medial," he added, tentatively. She nodded yes and reminded him that the medial nerves are medial to the axillary artery. When the student finished the demonstration, two of his teammates clapped lightly to encourage him. Marilyn ended with a few questions: "On the exam, what is the major landmark?" A female student replied, "The lateral thoracic artery." "Yes," Marilyn said, and added, "know your landmarks."

In this scene taken from my field notes, the TA, by presenting structures, teaches location and models pedagogical techniques, both of which are necessary for a successful demonstration. These predemonstration performances instruct students to recognize structures by their neighboring landmarks and relations and not just location, as location can differ from body to body. Students learn from TAs that these demonstrations involve tools (books, whiteboards, and instruments) and require active participants. Here students take on a teaching role as they guide their peers through the body by showing them anatomical structures. Ingold (2001) understands showing to be a "process of learning by guided discovery" (141), which instructs the novice to attend to what can be "seen, touched, or heard" (142), and thus can be made recognizable. This showing makes present the anatomical body while instructing other students in what is noteworthy, at least as far as that day's tasks are concerned.

Study Time

Students repeat these techniques of presentation during the afternoon, evening, and weekend independent study times. During these voluntary and unstructured sessions, students, usually in pairs or teams, walk around rehearsing their demonstration skills and anatomical knowledge learned through dissection and prosection demonstrations. The way TAs quiz students during the practice prosection is taken up by the presenting students who will, in turn, ask their classmates questions. These independent study sessions mimic the prosection demonstrations, only now students, in groups or pairs, take turns demonstrating and quizzing each other. The demonstrating students learn anatomy and learn to teach it by actively engaging students through questions, suggestions, and explanations. During formal class time and the study sessions, the presenting students must be prepared to answer (as best they can) their classmates' questions. Each team's practice demonstration with the TA at the beginning of class offers opportunities to ask the TA as many questions as possible. Knowledge—anatomical knowledge and pedagogical knowledge—passes from TAs to students. The students then pass it on to other students, perpetuating a specific language for the body. And this transmission of knowledge and know-how—and with it a certain view of the body—occurs through bodily practices of gesture, speech, and physical manipulation. These study sessions resemble those of the prosection course, save for the fact that medical and dental students have access to the cadavers during evenings and weekends.

Two Preparation Labs: "The Beehive" and the Future Surgeons

For students to engage in the embodied practices of anatomy education, the aesthetically sterile environment of the lab must transform into an intellectually stimulating, often entertaining, environment that incorporates a host of multimodal displays meant to index and supplement cadaveric bodies. To do this, the instructor enlists the help of TAs, a select group of applicants made up of former students of the course. In both courses, TAs, working in separate preparatory sessions supervised by the instructor, assist in setting up the lab space. The goal of these TA prep labs is to create an educational and representational environment in which students can learn human anatomy. Each anatomy laboratory course is, in actuality, two laboratories at once, with two assemblages of practices that share the same space—one to teach anatomy, the other to set up the displays for that teaching.

The Prosection Course Prep Lab: "The Beehive"

One of the anatomy instructors, William, aptly describes the weekly prep lab for the undergraduate course as "the beehive." During these sessions of intense collective work, 25 TAs prepare the labs for the 440 students enrolled in the 11 sections of the course. TAs, though still undergraduates themselves, assume explicitly pedagogical and professional roles usually reserved for graduate students. The instructor chooses them in part based on a prospective TA's letter of application sent to the instructor. The instructor offers no formal guidelines for these letters because he wants to see what potential TAs can do on their own. Looking for knowledge of anatomy, enthusiasm, maturity, and the ability to present information in a logical manner, the instructor needs strong TAs because they make the course possible:

> Well, my number one role is . . . to keep my TAs being good teachers and organizers. And it's my job to facilitate their role as a teacher and organizer. So for 440 students, my 25 TAs do the bulk of the work. I just facilitate their job and the material they are supposed to teach, the material they are supposed to teach with, and their, I guess, motivation sometimes. So if I had

25 terrible TAs, I'd had a terrible course. So the better I prepare my TAs, the better prepared they are to teach, and the better job they do. So that is why I prepare materials for the course. I mean the better the notes, the lab, the course packet, the easier it is for my TAs. The better the cadavers, the better the preparation. (William, instructor of both courses)

By training TAs in how to set up the lab space and prepare the resources (particularly cadavers), the instructor is also coaching these advanced undergraduates to think more like anatomists and anatomy teachers than like doctors or surgeons. In a formal training session before the semester begins, the instructor reviews university policies and course guidelines, talks to TAs about issues of assessment, offers tips on demonstrational teaching, and discusses the complexity of the TA-student relationship. After all, TAs will be teaching students who would be their peers in any other university context. This formal training helps TAs understand the values, perspectives, and responsibilities of teachers, thus initiating TAs into the particular orthodoxy of the anatomist.

During the prep lab, held each Friday afternoon from 3:00 to 6:00 p.m., TAs set up the space for the following week by dissecting cadavers, tagging structures, and composing the whiteboards. Like the undergraduate prosection lab, the prep lab begins with an agenda-setting meeting at the whiteboard, which involves organizing the tasks of the day and educating TAs on procedures and techniques. Because TAs perform the dissections required to prepare the lab, these TAs create the teaching prosections that students will use. The prelab lecture for the prep lab, conducted by the course instructor, reminds TAs what they should and should not do. The cadavers are the primary teaching tools and resources, so TAs have an important responsibility that begins even before students enter the lab—to create the key multimodal displays that are the origin of knowledge. By having newer TAs work alongside more experienced and advanced ones, the instructor allows them to mentor each other and learn through observation and collaboration.

The embodied practices and perceptual dispositions of these anatomists-in-training are collaboratively constructed and interactively conveyed as part of the practices of the prep labs. The TAs of the prosection course expressed to me a need to know anatomy and be mentors for students. One TA, John, understood it as his "duty to teach students everything that they're required to know, first of all, and, if they have interest, anything beyond that they wish to learn." The TAs view themselves as role models for students and less experienced TAs. In the words of another, Jason: "And when I'm in lab doing dissections, you know, I always make sure I'm prepared, that I know what I'm doing, and do my best, that way I can be an example to the other TAs." During my fieldwork, the TAs in the undergraduate course were the most interested in my study and curious about its pedagogical implications. Already viewing themselves as anatomists, specifically anatomy teachers, they wanted to know what my investigation's results might mean for their work with students and each other.

With music at times playing from a small radio on the counter, the atmosphere of the prep lab is relaxed yet serious. The instructor, who jokes more with the TAs than the students of the course, treats these apprentices more like colleagues than employees. After the instructor's initial directions, the TAs go about the business of dissecting and setting up. They first collect necessary tools, including forceps, scissors, tags, pipe cleaners, and string, plus the occasional scalpel, mallet, and chisel (figure 2.4).

As I discuss in the next chapter, one of the most important tools for dissection is Netter's *Atlas of Human Anatomy* (or *Netter*), which is the guide by which TAs identify a representative structure and dissect that structure so that students can later identify it. At this point, the instructor's supervisory role becomes more pronounced. Though he conducts dissections himself, much of his time is spent assisting and guiding TAs, offering second-order rules and coaching tips (Collins 2010). Preparing the lab also involves preparing the room's physical spaces and displays students will use to learn the structures on that week's ID list. The TAs set out the books, figures, posters, x-rays, and microscopes.

Again, another way for TAs to ready the lab is to ready themselves. All the TAs I interviewed mentioned reviewing the material and ID lists to make sure that they knew the anatomy, to hone their demonstration skills, and to prepare for student questions. The TAs, during the prep labs, aid and encourage each other, while modeling what some have described as a respect for the subject, the

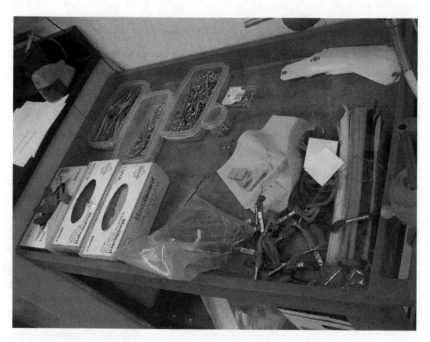

FIGURE 2.4 Common tools used in tagging anatomical structures.

instructor, and the course. As Jason put it, "I have to push them [the other TAs] to try to be the best they can be, so they also reflect nicely upon our professor." This particular laboratory, though much more loosely structured than the pro-section course, is nevertheless a space for learning anatomy by doing—in a more experienced TA's words, by "going above and beyond the ID lists." The instructor has organized this prep lab as a workspace in which an undergraduate class is prepared and teachers for that class are trained, developing advanced anatomical knowledge and a budding expertise through the practices of anatomical vision and their engagement with the multimodal displays that make these activities possible.

The Dissection Course Prep Lab: Future Surgeons at Work

Like the "beehive" prep lab (for the undergraduate course), the dissection prep lab (for the medical and dental course) begins with a prelab talk, one much more extensive and formal due to the advanced dissections these TAs perform. The dissection course covers more than twice the material of the prosection course, so TAs for this medical and dental course (13 third-year medical students and 1 fourth-year medical student) must cut more carefully to create what they themselves describe as "beautiful" dissections—clear, emblematic, 3-D examples (and instantiations) of the anatomical body. These formal talks almost always take place in the lounge close to the labs, where TAs sit at tables watching a professor discuss dissection techniques that take the form of second-order rules and coaching rules. Since these TAs have to dissect cadavers and then teach students using those bodies—all the while modeling how to present structures—prelab talks last about 30 minutes and cover not only what TAs should dissect but also how to help students dissect and demonstrate.

In the dissection course prep lab, TAs work two per lab room, meaning they often work without direct supervision from the course professors, although the professors periodically come through the labs to assist and answer questions. Usually TAs take turns: one will perform the prosection, the other will prepare the whiteboards and other resources. One TA might use *Netter* to draw out the pathways of the tarsal tunnel of the inner leg, including a common mnemonic for the structures: Tom, Dick, And Now Harry (or tibialis posterior muscle, flexor digitorum longus muscle, tibial artery, tibial nerve, and flexor hallucis longus muscle). Figure 2.5 shows a TA writing this mnemonic and a corresponding diagram on the whiteboard in his lab room. With *Netter* and the lab manual at hand, the TA composes a line drawing based on and faithful to the visual and verbal information presented in these course texts. Across the room, the other TA dissects these same structures on the prosection cadaver. Working together, the TAs create the anatomical objects and the physical learning environment.

FIGURE 2.5 A dissection course TA preparing a whiteboard.

The course professors expect these TAs to have mastered knowledge of anatomy and the practices of anatomy education—dissection, demonstration, and observation. Repeatedly, medical and dental students in the dissection course praised the TAs for their dissection skills. And many students admit to using the prosected cadavers as models to guide their own dissections because, as one female medical student said, "The experts did them. It is a look at how the body is supposed to look by someone who knows." Take for example, this male student's response to my question about the role of TAs:

> I think if your dissection is going rather poorly, and you're not finding anything, the TAs can usually come in and help, since they have a very large wealth of knowledge and experience. They usually know what is not important and what things you should be looking for. Also if you're being ultracautious, they can be somewhat cavalier because they know what they're doing; they know exactly what they're tearing through, you know, exactly what level to go through, and they can clean up stuff a lot faster than you can. (Ted, medical student)

TAs are perceived as having knowledge of anatomy and skilled capacities with dissection tools and techniques. Their understanding of the body, which includes knowledge of the cadaver's spatial and haptic dimensions, is the result of multiple

[handwritten margin notes: praise is part of learning how to appreciate the work of a "good" well crafted anatomical dissection]

years of study. TAs in the dissection course not only took the gross anatomy course but many also took an undergraduate course similar to ANAT 302.

This level of skill and socialization makes them excellent resources for examination tips and learning strategies, as they have gained an advanced anatomical knowledge and know-how and are currently deepening their technical expertise. In the words of Ted, a medical student: "Also if you're not finding a structure, they can point that out to you, maybe highlight some spatial relations that weren't discussed in the lecture portion." Though students can and do get such information from the instructors of the course, most students I interviewed relied more on TAs for study advice because students understood TAs as more like them: "I feel like a TA is the one you go to just to ask questions. They can give you, you know, the straight answer. You need to know this; you don't need to know this" (Amanda, dental student). As medical students and teachers, TAs are well versed in the practices required of both roles; also they are approachable sources of information on the social values and expectations of the course and medical and dental school more generally. As such, TAs serve as purveyors of embodied learning techniques and the cultural values and norms of the lab.

Perceptual Tools and Embodied Action

In anatomy labs, students and TAs enact embodied practices of learning that involve the creation of perceptual tools for the body and of the body. First, they use diverse tools and instruments (like forceps and scalpels) to transform the cadaveric body into a visual and multimodal demonstration (and instantiation) of the anatomical body. To dissect a cadaver is to reveal structures on and in the body. And by dissecting and revealing, participants transform the body into a 3D multimodal object designed to emphasize and communicate anatomical structures. Students and TAs inscribe these multimodal displays (such as atlas plates and x-rays) with anatomical knowledge by mapping anatomical terms and concepts onto these objects and then reading (or interpreting) them in light of that knowledge. For example, students navigate among displays of the anatomical body across a range of modalities—drawings of hearts, photos of hearts, x-rays of hearts, and actual cadaveric hearts. And they move among these displays and objects in order to make sense of the rich complexity of the body—specifically the heart—and to appreciate the variations that exist across specimen. These displays become perceptual tools of learning, shaping how students understand the heart.

Second, this creation of an artifact for the body is complemented by the ways in which the body is made into a perceptual tool. To learn, students and TAs use their own bodies as meaning-making instruments. Students walk around the lab to see different specimens from different angles. TAs and instructors teach students to use their hands as instruments of blunt dissection, to move, manipulate, and handle human tissue. Students learn to tactilely transform touch-based sensations of depth, texture, and pressure into information about the body—information they then convert into visual and, at times, mental representations of the body. All of this is done as part of the formation of a specific type of trained vision I call anatomical vision—or the perceptual, intellectual, and rhetorical framework by which participants make sense of anatomy. By developing this vision, participants acquire the embodied and perceptual dispositions that mark them as members of this medical community. These embodied practices instantiate the discourses of anatomy, render the physical body as the anatomical body, and persuade participants to understand the body according to a particular logic. Taken together, these practices constitute the work of the lab. Similarly, these physical practices—the movement around the cadavers, the physical manipulation of objects, the development of manual dexterity, the use of tools and texts, the performance of their own body in demonstration—make learning possible. Through embodied practices, participants in any domain acquire knowledge and skills (through formal education), while learning to enact a way of thinking and being based on cultural values (through informal enculturation).

Whatever our profession, the processes involved in constructing a perceptual tool of the body and for the body enact our formations of the world. And through the embodied practices involved in perceptual experience, we cultivate the trained vision facilitated by those instruments, objects, and discourses, a vision which in turn structures how we use these objects. Through their engagement with objects and instruments (such as forceps, cadavers, and their own bodies), anatomy students create perceptual tools that expand their cultural experience of the world, their knowledge of anatomy, and their understanding of the body. In technical environments, the objects and discourses we use—from mobile devices to technical vocabulary—not only expand our experience of the work we do and increase our possibilities for action, but they also expand our cognitive capacities and enable the world around us. In the words of Merleau-Ponty (1968), the world takes shape because of the "overlapping or encroachment" of bodies and objects—the ways our bodies couple with the objects, discourses, displays, and documents that surround us (123).

As researchers, teachers, and practitioners of TPC, we need to understand this entanglement of body-object-environment, as well as its implications, because the development of technical expertise involves the creation and deployment of perceptual tools (of and for the body) that make medical, scientific, and technical communication and education possible. To make sense of these embodied and rhetorical processes at the heart of technical work, I turn to phenomenological and cognitive theories of embodied action, specifically the enactive approach to cognition and the ideas of Merleau-Ponty that inspired that perspective.

How Embodied Action Enacts Meaning through Practice

Enactive theories of perception view cognitive processes as constitutive of the human being's capacity for action; thus, the mind is enacted through the physical body's activity in, with, and on objects of our environment. In one of the earliest articulations of this approach, Varela, Thompson, and Rosch (1991) assert that cognition "depends upon" the body's sensorimotor capacities, or its ability to move, touch, feel, and see. Thus human perception and, by extension, cognition are more than mental processes; they are forms of embodied action (173). For later enactivists like Noë (2006), perception is not a brain-bound mental process, but a "kind of skillful activity" accomplished by the entire body, not just the brain (2). Our perception of the world around us is determined by our bodily actions and capabilities, or (as Noë describes it) "what we do" and "what we are ready to do." Thus, we enact our "perceptual experience" through bodily activities, more specifically the sensorimotor capacities involved in seeing, touching, hearing, and moving (1). Rather than a computer that processes informational input from the environment, we more aptly resemble an organism that makes meaning through whole-body interactions with potentially meaningful objects of the world. This is true for even the most seemingly brain-bound activity, such as viewing objects.

Vision, Noë (2009) contends, is a skillful activity (60); more specifically, vision is a way of exploring the environment based on our understanding of sensorimotor knowledge that comes with guided activity (O'Regan and Noë 2001). For example, we understand that as we move around in a space, we will perceive objects differently based on our position relative to those objects; we know that our perception of an object will change as we move closer, or touch it, or raise it in front of us. Our visual experience of objects is similar to our touch-based experience of them, and visual experience "acquires content" through our "skillful movements" in relation to those objects (Noë 2006, 78). Vision, like all sensory modalities, is not something that happens to us; it is a form of action or set of actions we accomplish as we look around, focus our attention, and move ourselves to grasp (literally and perceptually) an object's properties (100). All experience, then, is a "mode of skillful encounter" with the world around us that enacts meaning because of our "mastery and exercise" of sensorimotor skills (231). As Varela explains (1992), perception is not a matter of receiving sensory input from "a pregiven world"; instead, perception and meaning making are processes that the perceiver enacts by engaging with a world of objects and subjects, of things and people (336). Our perceptual experience stems from the "continuous and reciprocal interactions" of "sensory, motor, and cognitive processes" (Thompson 2010, 256).

Simply put, perception is not something that happens to us; it is an activity that we engage in or enact. Cognition, or thought, is not solely contained in the brain, but is accomplished by way of the entire body. Perception and cognition are not only embodied, but they are embedded (in social situations of practice), extended (in that they often require other objects), and enacted (through our bodily coupling with the world around us). Menary (2010b) describes these points as "4E cognition" (459), while acknowledging that the theories and research at the heart of these approaches are not in complete agreement (see Rowlands 2010). An important precursor to these ideas of embodied, embedded, extended, and enacted perception—more specifically the enactive approach—is the phenomenology of Merleau-Ponty. Many philosophers of enaction draw heavily from Merleau-Ponty, adopting and adapting his theories of embodiment and perception. For Merleau-Ponty ([1945] 2005), all human actions, even those that seem socially and culturally derived, are embodied, in that there are no practices without bodies to perform and make sense of these actions. The human body is both our mode of expression and our capacity for action. In addition, Merleau-Ponty argues that we only know the world because we are embodied in the world—a world that comes to use through perception, which is always a bodily phenomenon (235–39). Specifically, the lived experience of being a body—interacting with the world's objects and discourses—allows us to perceive, judge, and know the world. Thus, we enact meaning through our interactions with the documents, displays, discourses, and objects of our environment.

The human body, then, is not only the text inscribed by human knowledge and practice, as it is when students map anatomical terms onto it. The body is also the vehicle through which we form this knowledge through practice, again as students do when they engage in skillful activities (like dissection) that allow them to enact the anatomical body. For Merleau-Ponty ([1945] 2005), the human body cultivates "our personal acts into stable dispositional tendencies" (169). These dispositional tendencies, which I understand as ways of seeing and thinking, are either played out or experienced by way of (he is never clear which) habit and habitual actions, which Noë (2009) describes as a type of "responsiveness to the environment" (125). Merleau-Ponty ([1945] 2005) goes so far as to say that it is the body, and perhaps not the conscious mind (or consciousness), that "understands" in the acquisition of habit, understands in this case being "the harmony between . . . the intention and the performance" (167). In other words, practices of habit formation and the dispositional tendencies we develop through embodied action inevitably navigate the middle ground between conscious experience and automatic, unconscious activity.

Through all of this, the human body "discloses the world for us" through the creation of perceptual tools borne out of these habits and dispositions, habits and dispositions that we use to experience that world (Merleau-Ponty [1945] 2005, 92). Often because "the meaning aimed at" by the body cannot be reached by its "natural means," the body must then "build itself an instrument" and "project thereby around itself a cultural world" (169). Though the ambiguity of the phrase "build itself an instrument" leaves unsettled whether the body makes an instrument *for* itself or an instrument *of* itself, I contend that Merleau-Ponty implies both. Our relationships with tools, instruments, objects, and discourses are always embodied relationships of incorporation and extension that create the cultural world around us. To illustrate, he provides the anecdote of the blind man who uses a walking stick to overcome or compensate for his visual impairment, thus altering his relationship to the world by granting his body a new perceptual tool: "Once the stick has become a familiar instrument, the world of feelable things recedes and now begins, not at the outer skin of the hand, but at the end

[margin handwritten note: dissection/an activity; a bodily action over another body; constructs anatomical knowledge.]

of the stick." Through his embodied use of this tool, as part of his lived experience, an experience that through habit and necessity contributes to his body's dispositional tendencies, the walking stick becomes "a bodily auxiliary, an extension of the bodily synthesis" (176). This prosthesis becomes incorporated into the phenomenological being of the body and the culture of the human subject.[1]

This bodily process of constructing a perceptual tool of the body and for the body is foundational to our formation of the world. Via this process we put into play the dispositions that the body makes possible through the deployment of tools, instruments, and objects that we use to experience the world. This process is a form of skillful activity through which meaning is enacted. Again, in the words of Merleau-Ponty, "We say that the body has understood and habit has been cultivated when it has absorbed a new meaning, and assimilated a fresh core of significance" ([1945] 2005, 169). And these instruments and tools, by which "new meaning" is absorbed into the body and the environment, need not be physical, material instruments. Discourses, objects, documents, and displays all involve skillful embodied activities to create and deploy. Through our engagement with them, we encounter new perceptual experiences and enact new meanings that shape our ways of seeing, thinking, and being.

Human perception, then, is a skillful activity "born" from our bodily engagements with the world (Merleau-Ponty 1968, 154). And like the objects that surround us, we are part of the world; we are enmeshed with the physical places we inhabit and perceive and the social spaces we mutually construct. Merleau-Ponty uses the term "flesh" to describe this intertwining of bodies and objects. Of course, he is not saying that humans and the world are not made of the same material, nor is he saying that inanimate objects have the same cognitive and sensorimotor capacities as humans. Rather, he is saying that we, like the objects of the world, make up (and participate in) the "flesh of the visible"—the common corporeality, the mutual encroachment that connects us to the natural and cultural environment (136). As a result, our perceptual experience, and thus our cognition, is rooted in bodily activities and sensorimotor capacities. We perceive, think, move, and feel through our whole bodily interactions and corporeal entanglements with the world around us. As a result, we are an assemblage made of bodies, objects, documents, discourses, and displays.

Perception is a skillful activity that involves much more than the brain. Our ability to think and to conceive of the world is dependent on our ability to experience it (Noë 2006, 208). Through our body's sensorimotor capacities, we learn to experience the world in meaningful ways. This process of meaning making is acquired through the development of bodily skills and habits, both of which are produced by our engagement with the environment, namely the objects, instruments, and discourses of that environment. If we think more specifically about a particular domain of practice, for example, the anatomy lab or the technical communication workplace, we come to see that perception and cognition in these domains are equally dependent on the body's skillful activities, the development of dispositions and habitual practices, and meaningful engagements with objects, discourses, documents, and displays that make up that world and shape our perceptual experience. Together with these objects, documents, displays, and discourses that surround us, we enact the social world by learning to engage with those objects in ways sanctioned by our cultures and our professions.

Embodied rhetorical action is the name I use for this process of social perception—what I might even call a program of social perception—through which much scientific, medical, and technical work is accomplished. Throughout *Rhetoric in the Flesh*, I develop this framework of embodied rhetorical action by reading my ethnographic data in light of phenomenological ideas of enacted perception and embodied action. Though I turn to these particular theories of embodied cognition, the impulse to uncover the mental processes involved in rhetoric and persuasion is nothing new. The induction to action and belief is a deeply rhetorical process that involves cognition, perception, and even emotion. Aristotle's examination of the emotions in book 2 of the *Rhetoric* provides one of the first systematic treatments of what we now understand as psychology (Kennedy 2007, 113). As Gross's (2006) history of emotion makes plain, thinkers from Descartes to Adam Smith often connected psychological processes and theories to rhetorical ones. Much of the animating force behind Burke's interest in rhetoric was his wish to understand human motivation (Burke [1945] 1969, [1950] 1969, [1954] 1984). In keeping with this, the field of cognitive rhetoric seeks to uncover the cognitive processes at the center of communication and persuasion, while also (subtly at times) emphasizing the historical interconnection among rhetorical, semiotic, and cognitive theories of tropes, figures, and narrative (Fahnestock 2005; Harris and Tolmie 2011; Lakoff and Johnson 1980; Oakley 1999, 2009; Turner 1996). This eclectic mix of rhetoricians, linguists, and cognitive scientists do this work by coupling rhetorical theory with cognitive linguistics and neuroscience. Recently, Jack and Appelbaum (2010) have identified a neurological turn in rhetorical studies, as scholars increasingly engage with neuroscience methods and findings. Jack and Appelbaum methodically map out the promise and peril, if you will, of these "neurorhetorics," which they see as involving both "the rhetoric of neuroscience" and the "neuroscience of rhetoric" (Jack and Appelbaum 2010, 412). The first approach analyzes the rhetorical formation of neuroscience claims and practices, while the second looks to neuroscience findings to offer insights into rhetorical and communication processes (412).

While my approach to phenomenological and cognitive models of enaction more closely resemble Jack and Appelbaum's second category, I read and use these ideas cautiously, fully aware that these theories are rhetorical artifacts that are, in Gross's words, "designed to persuade" (1996, 5). For example, these phenomenological theories of enaction are often presented as rebuttals of more computational theories of mind, which view the brain as a type of computer that processes sensory information from the outside world. To engage with these theories, I must inevitably (though reluctantly) weigh in on contentious debates in philosophy and cognitive science. One such debate involves whether perception (and thus cognition) requires the creation of mental representations or mental images of the world in order to think and recognize objects. Enactivists such as Noë (2006, 2009) and Thompson (2010) reject the mental-model hypothesis espoused by other cognitive scientists such as Kosslyn, Thompson, and Ganis (2006). The mental-model, or mental-representation, hypothesis contends that perceptual experience and thus cognition requires mental models of the world that contain semantic properties we tap into in order to make sense of the world (Thompson 2010, 25). Enactivists, conversely, view perception as direct in that we do not, as the computational theory describes, build up internal models by which we filter perceptual experience (Noë 2006, 22; Thompson 2010, 25). This notion of direct perception has perhaps as many critics (Adams and Aizawa 2008; Block 2005; Rupert 2009) as adherents (Chemero 2009; Hutto and Myin 2013; Stewart, Gapenne, and Di Paolo 2010). There are even those who reject the mental-representation hypothesis but disagree with the explanation enactivism provides (Pylyshyn 2001, 2003). At the same time, other cognitive scientists and philosophers are adopting and adapting the enactive approach to shed light on such phenomena as social interaction and social cognition (De Jaegher 2009; De Jaegher and Di Paolo 2007; Fuchs and De Jaegher 2009). Recently, Hutchins (2010) has reframed his concept of distributed cognition as a form of embodied enaction, whereby humans use resources in the environment not only to distribute the cognitive load and accomplish cognitive tasks, but also to enact meaning through skillful interactions with these objects and people. (I return to this concept in chapter 5.)

As an empirical researcher, my entry into the mental imagery debate is influenced by my data (interviews, field notes, photographic documentation, and course materials) and my modes of analysis (interpretative methods of qualitative coding). Because I observed participants engaged in situated practices in an educational setting, rather than the fMRI (functional magnetic resonance imaging) results of participants engaged in simulated actions, I cannot judge what was happening inside their brains. I do know that for participants in the gross labs, visualizing (forming) what we take to be mental images is an important component of learning anatomy. As I discuss in the next chapter, these visualizations of anatomical structures and anatomical bodies are created by participants' physical interactions with lab's multimodal displays. These mental images that participants rely on are formed from their ability to remember the various visual displays. Their ability to visualize the body is predicated on their interactions and familiarity with the various multimodal objects, which they later remember through visualization. In other words, when participants talk about visualizing anatomy, they are actually talking about remembering. And through sensorimotor capacities, namely seeing, moving, touching and, at times, hearing, participants learn to make sense of these images and objects, calling them to mind later as they seek to make sense of other displays. In this particular domain of technical work, embodied actions make these seemingly mental perceptions possible.

Conclusion

Anatomical education involves rendering, materializing, and making visible the anatomical body on objects that often resist a transparent rendering, on bodies we already seem to know so well. Learning anatomy is relearning configurations of the body, along with new values and attitudes. Gradually, over time, participants accomplish this reconceptualization, this trained vision, through a kind of skillful interaction with the objects that surround them. Specifically, students, TAs, and instructors inscribe these objects with the discourses, terms, and concepts of anatomy through the embodied practices of looking, touching, moving, cutting, speaking, and making. And they incorporate these renderings into their conceptualization of the body, in the process shaping how they engage the abstract, conceptual body of medicine and the physical body of lived experience. The work of the gross laboratory is constructed around an assemblage of objects, bodies, and discourses that "make" the anatomical body through anatomical vision. This body-object-environment assemblage, what I call embodied rhetorical action, enacts a physical, perceptual network for the body and of the body, one that operates to form a particular embodied socialization that will structure their attitudes, judgments, and future actions.

To make sense of these social and embodied practices, I turn to phenomenological theories of embodied mind and rhetorical theories of demonstration and display. I make this turn at a time when an increasing number of humanities and social science disciplines are looking to the sometimes unquestioned hypothesizes of neuroscience and cognitive science as a means of (understandably) enriching our work and (more disconcertingly) legitimating what we do by borrowing from a more scientific discipline (for example, Connolly 2002; Massumi 2002). Let me be clear that my wish is not to bolster some inherent weakness of rhetorical studies and TPC with the rigor and prestige of cognitive science. Rather, based on my fieldwork, my analysis of data, and my personal experiences as an embodied human in a gross lab, I am convinced that an approach to situated practice that couples rhetorical theory and theories of enaction offers original insights into communicative activity, specifically the development and coconstruction of technical expertise and trained visions. Recently, Rivers (2011) has provocatively posited TPC research as "a necessary complement to cognitive science" (413). He finds in our work with disciplinary tools, workplace environments, and performance assessment points of potential convergence with, in particular, Clark's (2011) cognitive theories of extended mind. No doubt the work of Swarts (2008), Van Ittersum (2009), Winsor (2003), and others attests to the field's potential as a rich thought space for notions of distributed and extended cognition. And I agree with River's contention that TPC is uniquely suited to contribute to theoretical and practical conceptions of minds, bodies, perception, and cognition. More specifically, I am convinced that a rhetorical approach to enactive cognition provides a robust and portable framework for explaining and enriching situated practices. *Rhetoric in the Flesh* is in part evidence of that conviction.

3

LOOKING AT PICTURES

Multimodal Displays and Perceived Affordances

The visual displays of the gross lab contribute to student learning as aids to dissection, objects for demonstration, and visualization tools for observation. Rhetorically speaking, anatomical displays are demonstrations that exhibit or manifest anatomy, operating as proof or evidence of a particular anatomical structure. Whether containing human material or representing such material, these multimodal demonstrations make anatomical knowledge visible and legible in a way that participants can deploy as they engage in the work of learning. For centuries, learning anatomy has meant learning to make sense of visually dominant multimodal displays, either representational images or human bodies exhibited to demonstrate anatomical knowledge and evidence (Cunningham 1997). Even during the supposed revolution in early modern anatomical science that encouraged direct observation and human dissection, such 2-D visual displays as anatomical illustrations were important demonstration tools (Carlino [1994] 1999; French 1999). At least since the sixteenth century, anatomical demonstrations have required coordinated interactions between living humans and multimodal displays.

Figure 3.1 is an illustration from Banister's (ca. 1580) *Anatomical Table*, depicting a dissection presided over by the anatomist himself. Before an audience of students and professors, Banister demonstrates structures in the human abdomen. With a pointer, he compares the cadaver on the table with the skeleton beside it. The anatomy book (Colombo's *De re anatomica* of 1562) rests open between him and the skeleton.

With one hand in the viscera of the abdomen and the other indicating to the corresponding location on another body, Banister relates the view provided by one display (the open cadaver) with that provided by another (the skeleton). The book—along with the anatomical knowledge it represents—mediates his view

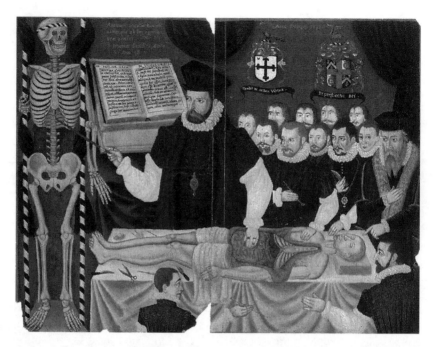

FIGURE 3.1 Frontispiece of Banister's *The Anatomical Table* (ca. 1580). Reprinted with permission of University of Glasgow Library, Special Collections.

of that skeleton. By way of visual parallelism, the two pages of the open book, the two sides of the skeleton's ribcage, and the two flaps of reflected skin on the cadaver all seem to mirror each other. Like anatomists before him, Banister, with this illustration, seems to suggest that the knowledge of the open book can be viewed in the open body if we are trained to recognize it. Carlino ([1994] 1999) reads in this image proof of the Renaissance's anatomical practice of connecting "knowledge acquired from books" with "empirical observation acquired from teaching" (59). What Banister's image also makes plain, I contend, is that demonstrating anatomy involves not only cadavers but also other visual (and at times verbal) displays used to bring anatomy before the eyes. This point is as true today as it was then, in large part because different displays provide not only different views of the physical body but also different opportunities for making sense of that anatomical body on display. Students today, as in the past, learn to recognize the anatomical body in these images and objects.

Anatomists rely on these multimodal displays to fulfill what Van Dijck (2005) describes as the "two contradictory requirements" of anatomical pedagogical objects, namely examples that provide both "authenticity and didactic value" (43). These displays, whether cadavers, wax models, or photographs, must offer authentic, lifelike representations as well as comprehensible, instructional usefulness. That is, anatomical displays must be purposefully rendered to make visible anatomical structures as they are typically found in the body (authentic value), yet these displays must supply clear pedagogical examples that can teach valuable relationships (didactic value). By demonstrating structures in an actual body, a cadaveric dissection revealing the uterus and uterine tubes (fallopian tubes) offers an undeniable authenticity. But because of the uterus's position in the body and the neighboring organs that crowd it, a student viewing (and even touching) this display might not be able to locate the fimbriae and ovaries. An atlas illustration or a line drawing of the uterus and uterine tubes, while in no way an authentic example, does offer a clearer view of the location of the fimbriae and the ovaries, not to mention information about their connected structure–function relationship. The 2-D displays of atlas plates and drawings have real pedagogical value that assist students in learning anatomical structures and relationships as well as in making sense of cadavers.

To represent and present anatomy, teachers and students must find ways to reconcile or supplement authenticity and didacticism. Most commonly, they achieve this by displaying and using an array of visuals: drawings, atlas illustrations, photographs, radiographs, histological images, plastic models, cadavers, and other medical images such a PET scans and CT scans. Because no single display can exhibit all bodily features at once, different objects provide different vantage points by affording different ways of knowing and experiencing anatomical bodies and knowledge. For instance, a teacher might use a display depicting the complexity of the cadaveric body (such as a photograph) in conjunction with a more stylized and simplified display (such as a plate from *Netter*) that might better elucidate a pedagogical point or structure–function relationship. The lab's visual and multimodal displays, then, carry different affordances and opportunities that make them useful and meaningful in varying ways. To demonstrate anatomy, to bring anatomy before the eyes, participants must learn to accurately read, so to speak, the picture the display provides as well as recognize the didactic and authentic usefulness of differing displays and objects.

In this chapter, I analyze the (physical) apodeictic displays of the anatomy lab by attending to the objects and practices that make possible anatomical demonstration. I introduce and classify the networks of visual and multimodal objects. I do so to demonstrate how participants use and build meaning from these displays by recognizing and enacting the objects' solicitations to act, or what Gibson (1986) terms perceived affordances. Engaged in the interactional processes of demonstrating anatomy, participants conceive of the lab's visually oriented multimodal objects as material instantiations of the anatomical body as well as opportunities to engage in skillful action. This simultaneous recognition and enactment of meaning is dependent on each display's representational or presentational content, dimensionality, materiality, and perceived level of interactivity. Through an exploration of the anatomy lab, this chapter explains how multimodal objects, in any domain of practice, carry different meanings, in part because they allow different opportunities for action. Participants perceive these opportunities for action through skillful engagement and the budding expertise that shapes how they view those objects and actions. These multimodal displays and the embodied rhetorical practices they encourage ultimately configure a trained perspective based on the enacted affordances of objects on display. Yet trained vision is not only developed by way of practice; trained vision and practice mutually constitute one another. And this dual articulation becomes most apparent not through a decontextualized analysis of multimodal displays but through direct observations of displays used in comparative, rhetorical actions.

Multimodal Displays as Tools for Action

One morning in the dissection course (ANAT 600), I stood among the students, who were assembled in their body-buddy teams around the table-like cadaver tanks. Together we watched the TA Brian, poised before a large whiteboard filled with diagrams and reminders, ready to give that session's prelab talk. The students and the other TA, Ruth, stood close by and directed their attention to Brian and the board. Beginning with dissection tips, he reminded students to use "the lab manuals, *Netter*, and *Grants*" as they dissect. As the course's primary dissection aids, these texts—Netter's *Atlas of Anatomy* and *Grant's Dissector*, along with the instructor-created lab manual containing tips, mnemonics, and ID lists—were indispensable to students and TAs alike. At the end of Monday's lab, students needed to be able to physically draw the nerves and veins very well. "This activity will help you learn," he added. "The lateral, posterior, and medial nerves are the more obvious ones, but you will need to know the others as well."

Then, Brian directed the students to pages in *Netter*. "OK, let's go to plates 443 and 447," he told them, turning pages of the book that rested on the easel in front of him. "The point is not to wait until the end of the upper extremity to discover these plates. They are very helpful." He guided them to plate 169, stating that they should be able to draw a typical spinal nerve. He held up the book to show them the illustration, directing one team to keep their finger on that page. "It is like a phone cord. It is dozens of fibers," he told them, describing what is not easily apparent from the *Netter* plate. "You open up the nerves, and there are dozens of fibers." Then he moved from *Netter* to the whiteboard behind him, pointing out key dissection techniques, mnemonics, and the lists of parts to be identified during dissection. At one point, he walked over to the door, indicating with his finger a color copy of a photograph hanging there. The image, from Gunter van Hagens's *Body Worlds* exhibit, displayed a plastinated cadaver seated with his (or her) back to the camera. Brian directed the class to the photograph's demonstration of the ventral and dorsal rami (divisions of the spinal nerves). "You should look for these in your cadavers." He admitted, however, that dorsal rami are so small they might not see them.

This scene, adapted from my field notes, provides an emblematic example of what students, TAs, and instructors encounter in a gross anatomy lab—an assemblage of visual displays, physical objects, and material practices that make their work possible. Brian directed students to the course's major textbooks: Netter's *Atlas of Anatomy*, which offers a host of visual representations of the dissected body, and *Grant's Dissector*, which offers instructions, explanations, and illustrations to guide dissection. Participants described these two texts, one image dominant and the other word dominant, as "road maps" for the course, because they offered examples of how the dissected cadaver should look (*Netter*) and explanations of how one accomplishes this (*Grant's*). To arrive at the well-dissected cadaver and to learn anatomy from cadaveric bodies, students must navigate a series of multimodal objects, moving physically back and forth between cadavers, visual displays, and other course texts. Not unlike Banister's sixteenth-century illustration, students, TAs, and instructors in the contemporary lab use these maps not just to find their way, but also to create the path they follow and the final destination they are aiming toward.

Dissection is also a process of navigation in which participants learn anatomical structures by excavating and exposing layers of human tissue, thus transforming the body into a multimodal illustration. Dissecting to reveal the dorsal rami, for example, changes the cadaver into a 3-D visual illustration of those structures. But this visual and physical alteration is not always possible. The cadaver, after all, has its limitations; it is dead tissue, subject to drying out (due to exposure) and evisceration (a result of dissection). Because of this, Brian directed the class to an illustration of *Body Worlds*'s "Chess Player," which exemplified structures by showing them in a real human body and offering subtle clues concerning the structure's physiological purpose. By looking at the "Chess Player," students can see these particular nerves in action (so to speak) and in context (in an actual body); thus, students can understand where nerves are located and deduce information about function.

This scene also exemplifies the way multimodal communicative acts are constituted through what Norris (2004b) terms "modal density," or the dynamic interplay of multimodal objects that seem to operate simultaneously (103). In Norris's (2004a) example of a woman shopping in a grocery store, the multimodality around her consists of visual, aural, spatial, verbal, and gestural modes from which she draws meaning. The meaning she derives, I argue, is dependent not only on the modes deployed and her interpretation of those modes and objects, but also on the skillful activities she uses to interact with those objects. Norris's (2004a, 2004b) work illustrates Kress and van Leeuwen's (1996, 2001) contention that all communication is unavoidably multimodal, an assertion that encourages investigation of the modes and meanings used in communicative acts (Kress and van Leeuwen 2001; Kress 2005). This inquiry has led to research into the interpretation of multimodal objects (Ball 2004; Bezemer and Kress 2008; Hull and Nelson 2005; Wysocki 2001), the composition of such objects (Anderson 2008; Fraiberg 2010; George 2002; Wysocki et al. 2004) as well as possible means of assessing them (Katz and Odell 2012; Potts 2009). Much of this work in multimodal analysis provides insights into what Maier, Kampf, and Kastberg

(2007) describe as an object's "interconnectivity of modes" and modal dominance (454), which researchers uncover through a "fine-grained analysis" of an object's "visual and textual interactions" (457). Yet all of us move and take part in a world of multimodal objects, at times without consciously reflecting on this modal density. We interact with objects and displays through an ongoing and seemingly automatic analysis of their modal dominance and modal interactions. And I contend that we do this as part of our perceptual experience of attending to objects while we engage in some intentional (or world-directed) task such as shopping, touring a museum, or learning in a classroom.

In the gross anatomy lab, participants engage with the modal density around them as part of their ongoing activities of learning anatomy, specifically demonstration, observation, and dissection. Brian pointed students to different objects and displays because they offered different vantage points for viewing and understanding the anatomical body. Students navigate a modally dense environment by learning to appreciate and recognize the way objects and displays signify differently depending on their modal prominence. For example, the use of color and layout in a more detailed display such as a *Netter* plate offers a didactic value that can be obtained faster than a verbally dominant written description. The *Netter* plate in figure 3.2 displays key anatomical structures that medical and dental students dissected during the first lab session.

The illustration in figure 3.2, which an instructor showed during the first day's lecture, offers a clear picture of the cephalic vein, one of the structures students

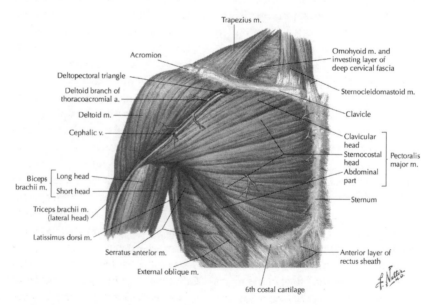

FIGURE 3.2 Plate from Netter's *Atlas* depicting the pectoral region. Reprinted with permission from *Netter Images*.

had to identify. Phenomenologically speaking, this display is a representation because it "re-presents" another object; it furnishes a pictorial image (a view of the body) that represents a pictorial subject (a real human body). Here I borrow from Husserl's (2006) identification of the three components of picture viewing or picture making: (1) the "pictorial vehicle," or the actual physical object; (2) the "pictorial image," or the image (or resemblance) that is represented in or through the pictorial vehicle; and (3) "the pictorial subject," which is the referent the pictorial image indexes (also see Thompson 2010, 288). Viewing this illustration, or any image, is a quite complicated endeavor because we must identify this pictorial image as a body, a recognition gained through formal training and informal socialization. Of course, a labeled atlas illustration such as this does make recognizing, identifying, and learning much easier. This idealized representation (free of fat and fasciae), with its use of labels and color-coordination (muscles and arteries are depicted in red, the veins in blue), provides the didactic value of this anatomical display. Students and TAs use this pictorial image to recognize and learn more about the pictorial subject, in the process emphasizing the cephalic vein's location to the deltoid branch of the thoracoacromial artery.

A photograph of an expert prosection, on the other hand, offers an authentic value that an idealized atlas cannot, yet that authenticity can make visual identification challenging. The photograph in figure 3.3 was also displayed during the first lecture just before students carried out their first dissections. Here, we see the same region of the body depicted in figure 3.2.

The instructors, who staged and took this photograph, have labeled the cephalic vein and the deltoid branch of thoracoacromial artery for easier identification. Like the *Netter* plate, this photograph depicts the same pictorial subject, this time displayed in the monochromatic cadaveric body. This image is representational also, but it offers more realistic detail, perhaps a gruesome realism, that the image in figure 3.2 does not. In both, labeling makes identification easier. However, in real cadavers, those labels are absent, so students must use these pictorial images to learn the structures (pictorial subjects) well enough to identify them in any multimodal displays (pictorial vehicles), whether cadavers, photographs, or atlas plates. Learning to recognize accurate pictorial images in such photographs or atlas illustrations takes training and repeated exposure. Students often read one display in light of another in order to reconcile the pictorial images that both provide. Learning to see in a gross lab involves learning to enact the affordances that multimodal objects (pictorial vehicles) seem to provide, affordances dependent on the actions of participants. Students might prefer the photograph in figure 3.3 over the *Netter* plate in figure 3.2 when seeking to identify the deltoid artery, for example, in a cadaveric body. Based on their anatomical value and their dimensionality, materiality, and perceived level of interactivity, multimodal displays in the gross lab seem to invite certain actions. These "solicitations to act," in Dreyfus's (2005) words, are what the lab's displays seem to provide. Even outside the lab, we make sense of multimodal displays and

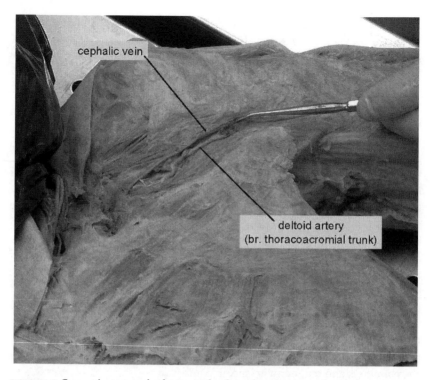

cephalic vein

deltoid artery
(br. thoracoacromial trunk)

FIGURE 3.3 Course instructor's photograph of a cadaver's pectoral region.

perceive them as offering opportunities for action as we engage in the skillful activities of enactive cognition.

Affordances as Opportunities for Action

To articulate more precisely how our interactions with multimodal displays encourage perceptions and actions, I turn to Gibson's (1986) notion of affordances. As part of his "ecological approach" to visual perception, Gibson describes affordances as what "the environment" offers "the animal," that is, "what [the environment] provides or furnishes, either for good or ill" (127). Affordances are "perceivable and are usually perceivable directly" because they are what the physical world offers up to the organisms who inhabit that world (143). An object's affordances are part of an individual's experiential, embodied, and lived-through (in the Merleau-Pontyan sense) encounters with the world. In other words, affordances are "opportunities for action in the environment of an organism" (Sanders 1999, 129).

Gibson (1986) provides the memorable illustration of how a surface such as a ledge or tabletop can afford sitting if it is "horizontal, flat, extended, rigid, and knee-high relative to the perceiver" (128). Affordances, or opportunities in the environment, are visually perceived and rooted in embodied perception, meaning they are contingent upon an individual's visual, bodily, and spatial orientations. An affordance is "equally a fact of the environment and a fact of behavior," pointing in the direction of both "the environment" and "the observer" (129). Affordances depend on both the environment and our perception because our "awareness of the world" and our "complementary relations to the world" are inseparable (141). An affordance is a circumstance of our embodied, perceptual movement through the complex terrain around us. (Gibson is unclear as to whether an object's affordances are a cause or a result of our embodied practices.) In language reminiscent of Merleau-Ponty ([1945] 2005), Gibson (1986) understands the human being (or organism) as "created by the world" he or she inhabits; as such, an individual fits "into the substances of the environment" and is "formed" by that environment and that act of fitting (130).

Arguably Gibson's concept of affordances seems to describe some essential nature of objects, of things-in-themselves. After all, an object's affordance, for Gibson, "does *not change* as the need of the observer changes" (1986, 138–39, italics in original): "An affordance is not bestowed upon an object by a need of an observer and his act of perceiving it. The object offers what it does because it is what it is" (139). This description of the object seems to ascribe agency to things (which may or may not be productive), and more problematically, it bestows a timelessness and autonomy to the object that the organism (constituted and formed by the world) does not seem to experience. Sanders (1999) stresses not the permanence of the affordance but its temporal consistency, meaning that an object's affordances do not change "as the moment-to-moment needs or moods of the observer change" (129). Gibson (1986) views the environment not as a meaningless landscape on which the subject imposes meaning, but instead as one already full of the capacity for meaning (meaning-full-ness, if you will), which is perceived and fulfilled by the human who is founded in the perceptual world. Affordances of the world are not a "value-free" series of options but rather a "value-rich" ecology that we encounter as we negotiate our environment (140). Again, Merleau-Ponty (1968) conceives of our relationship to objects as a reciprocal overlapping of body-object-environment that constitutes what he calls "the flesh of the world" (136), or an intertwining of visible, tangible objects (and bodies) that are full of the potential for meaning (134).

Noë (2006), who reformulates Gibson's ideas in light of enactive cognition, asserts that we experience the properties of an object as we "grasp" that object's "sensorimotor profile." In the process, we experience that object as determining possibilities for movement and action. And this experience of the object is an unavoidable component of perception, our way of understanding how the world around us structures possibilities for action. An object's supposed solicitations to act, then, are contingent upon the animal's bodily capacities and skills (106). An affordance, I contend, is neither solely in (or with) the object nor solely in the mind or body of the user; rather, affordances are born, so to speak, from the merger of object and body. It is through this intertwining that we learn to (at once) invent and discover these opportunities according to the dispositions into which we are socialized and the rhetorical and cultural situations in which we find ourselves. We enact affordances through our skillful engagement with objects, yet this enactment is structured by larger social and cultural influences.

Returning to the practices of the gross lab, participants perceive multimodal displays and objects as conveying particular affordances, and participants take up these opportunities for action as they navigate the spaces and engage the objects of the lab. These affordances shape not only how students use these objects, but also how students incorporate these objects (and their affordances) into their learning—their trained vision—which mutually influences and makes possible the embodied practices. The enacted affordances of a multimodal object are dependent on that object's properties, namely its dimensionality, materiality (including media), level of interactivity, and representational content (or level of mediation). In fact, an object's supposed usefulness and usability is structured by the opportunities for action these properties allow. For example, Becky, an undergraduate, explains how the whiteboards were often of little use to her:

> I don't generally use it [the whiteboards] unless I see something that really sticks out as, like, "O.K. That is helpful" or "O.K. That is something I am probably going to need to identify, so I need to know that." So I write that down. Otherwise, I don't really need the whiteboards or use them. I am more of the visual person. I need to actually see it and play with it rather than just read it.

Here Becky dismisses the whiteboards as unhelpful because they do not fit with her perceived visual style of learning. She only uses them when something "really sticks out" at her. In Gibson's terms, Becky uses the whiteboards when she perceives a meaningful affordance. What "sticks out" is information she deems helpful or necessary to know. Here we see how the larger goals of a community (in this case, the course's goals, objectives, and exams) influence an object's perceived affordances. And when the whiteboard information affords taking note of, Becky does; when it does not, she does not. I make this point to illustrate, first, how a multimodal display's enacted affordances not only dictate its usefulness but also shape how participants perceive it in general and, second, how our larger communities of practice shape our enaction of affordances. Becky's integration (or perhaps conflation) of visual learning with kinesthetic learning is worth noting. She describes herself as a "visual person" who needs to "see" and "play with" an object, and not just "read it." Though the whiteboard, full of line drawings and descriptions, is a flat, 2-D space, Becky perceives of its potential usefulness in tactile, 3-D terms. Its usefulness "sticks out" as "helpful," which she interprets as a solicitation to act, namely to "write that [information] down." These metaphors of embodied engagement perhaps stem from the fact that she will use this information during her physical interactions with cadaveric bodies.

As Becky's words suggest, culture and socialization play a significant role in our ability to enact an object's affordances. Wartofsky (1979) contends that our visual ecology is "cultural" and "sociohistorical," not "natural" or biological (275). Our decision to sit in one section of a chair and not another, after all, has less to do with the affordances of the chair, which offers both the seat and arm as sit-on-able, and more to do with the social context and cultural dispositions into which we are socialized. Our actions are structured by the goals, norms, and values of the groups to which we belong. Those groups authorize as legitimate certain actions while foreclosing the possibility of others, which Bourdieu (1980) describes as a consequence of our habitus. Our ability to make sense of those objects and to enact their affordances is shaped by our social group. In the gross lab, determining whether or not an image or object is a representation can have serious pedagogical consequences. The anatomical value of a display derives from its (perceived) authenticity or didacticism, which often relates back to the display's perceived level of mediation—specifically whether the object represents (or refers to) anatomical structures or presents anatomical structures (by being or containing human material). This mediation is enacted through the perceived affordances of the object, which are constructed by the object, our actions, and the values and norms of anatomical science.

Even learning to recognize the pictorial image a visual display exhibits is a process of enaction dependent on our skillful encounters and larger socialization. As Hutchins (2010) reminds us, the ability to recognize an image, or "to apprehend a material pattern as a representation" (429) is determined by "culturally shaped perceptual processes" (429–30). The perceptual skills allowing us to recognize images are "learned cultural skills" that we exercise through skillful engagement with those objects (446). Through these interactions, we, in Wartofsky's (1979) words, "create similarity" and "likeness" in pictorial images by recognizing the likeness or assuming the likeness is present (277). What counts as a representational display, then, has less to do with a representation's (or pictorial image's) fidelity to the original referent (or pictorial subject), and everything to do with our perceptual experience of the image (Wollheim 1965, 9). This perceptual experience is shaped by cultural practices. We enact representational qualities of images by perceiving mimetic or metonymic relationships. Although an object, again visual or otherwise, has physical features and conventions that seem to imply these actions and make it perceivable as soliciting these actions, these affordances are not as much *in* or *of* the object (50). Instead visual and embodied opportunities for action are formed through the engagement of the embodied spectator and the object through skillful interactions formed by and between body and object. The body-object-environment assemblage enacts affordances.

Our culturally and socially derived abilities to interpret and assess are important components of enacting an object's affordances. Affordances stem in part from, as Collins (2010) describes, "the interpretive capacities and tendencies" of the human being interacting with an object (39). Affordances make actions possible only for users who can enact those possibilities by recognizing and seizing what is present. This enactment happens as users interpret an object's features or conventions according to frameworks provided by their community. Kostelnick and Hassett (2003) use Gibson to explain how technical communicators interpret and use visual conventions; as they contend, affordances are "the visual elements" that offer an audience "the opportunity for interpretation" (224). Enacting (recognizing and seizing) an object's affordances means seeing its modal elements and object properties as "offering opportunities for meaning" and interpretation (225). A display's affordance, which we enact by recognizing what is present, makes interpretation possible. Our interpretation and use of displays depends on their object properties, namely modal density, materiality, dimensionality, interactivity, and mediation, as well as our rhetorical purposes for using them. Our awareness, even perception, of these properties is filtered through, yet mutually constructed by, our trained vision. In short, we perceive, interpret, and use multimodal objects based on our embodied perceptual experience of the objects' affordances, our socialization, and our rhetorical purpose.

Multimodal Displays of the Lab

In the anatomy lab, diverse and complementary multimodal displays make possible the rhetorical and embodied practices of learning, teaching, and communicating anatomy. Participants use these displays to learn anatomical knowledge by mapping that knowledge (and the anatomical body) onto the various 2-D and 3-D displays that either represent the anatomical body or present some version of an actual body. Again, to use these displays means perceiving and seizing the opportunities for action that these displays seem to afford, affordances we enact through our engagement with these objects. Specifically, the recognition of didactic and authentic usefulness invites participants to interact with them in the performance of actions that the displays and participants collaboratively make possible. These invitational actions are dependent on and perceivable by way of the display's dimensionality, materiality, perceived levels of interactivity and mediation, along with the goals of the course and the values of anatomical science.

Based on my analysis of ethnographic data, I divide these multimodal displays into three broad and somewhat overlapping categories based primarily on their anatomical value: (1) didactic representations, (2) authentic representations, and (3) authentic presentations. The first category, didactic representations, is made up of 2-D images and 3-D objects that are absent of biological material yet enact the anatomical body by representing or referencing actual bodies. The second group contains representations derived from medical imaging technology that also lack biological material; however, participants perceive these 2-D images as offering a unique and authentic access to the body's interior. The final group, authentic presentations, consists of displays containing human material; thus, participants understand these as offering presentations of real anatomy. Except for the 2-D microscope slides, most of the displays in this third group are 3-D objects. In what follows, I classify each major visually and tactilely oriented multimodal display of the lab according to these three categories. I do this to illustrate the role dimensionality, materiality, interactivity, and mediation play in the bodily enactment of affordances; see tables 3.1, 3.2, and 3.3 for an overview of these displays. In the rest of this chapter, I use the term "image" to refer to 2-D (or seemingly 2-D) displays, and the term "object" in reference to 3-D displays.

TABLE 3.1 Didactic representations: Visual displays absent of biological material, representing (or referencing) "real" anatomical structures.

Visual and Verbal Schematics (Didactic, 2-D, Low Interactivity)	1. Line drawings 2. Verbal diagrams
Naturalistic Images (Didactic, 2-D, Low Interactivity)	1. Color illustrations 2. Atlas plates 3. Wall posters 4. Photographs
Nonhuman Corporeal Objects (Didactic, 3-D, High Interactivity)	1. Plastic models

TABLE 3.2 Authentic representations: Visual displays absent of biological material, offering unique access to the body's interior through medical imaging technology.

Nonhuman Corporeal Displays (Authentic, 2-D, Low Interactivity)	1. Radiographic images (x-rays) 2. Advanced imaging (CT or PET)

TABLE 3.3 Authentic presentations: Visual displays made from biological material, presenting "real," authentic anatomy.

Human Corporeal Displays (Authentic, 2-D, Low Interactivity)	Histological images (microscopy)
Human Corporeal Objects (Authentic, 3-D, High Interactivity)	1. Plastinated body parts 2. Excised body parts 3. Cadavers 4. Living Bodies

Visual/Verbal Schematics: Didactic Representations I

The first group of displays comprises the simplest texts imaginable, some of which hardly seem like images at all. Take, for example, the whiteboard in figure 3.4, where a large line drawing of what is meant to be a (perhaps male) human body demonstrates the "Cutaneous Innervation of Upper Extremity." Colored lines on the arms and the corresponding legend to the right instruct the viewer on the systems of innervation that make motor function possible.

To use this schematic drawing, students must read the image by way of the color-coded interactions between visual and verbal elements and verbal interactions that offer them a pictorial explanation for the relationship between nerves and muscles in this region of the body. Similar informational graphs on the whiteboards in both courses form most of the visual and verbal schematics. I call them schematics to emphasize how these image-based displays (usually line drawings) and word-based displays (usually verbal mnemonics) provide a rudimentary understanding of anatomical structures and relationships, primarily signifying through the interaction of visual and verbal elements. Also, my term is inspired in part by the terminology of anatomical atlases such as *Gray's Anatomy* that describe simplistic drawings as schematics.

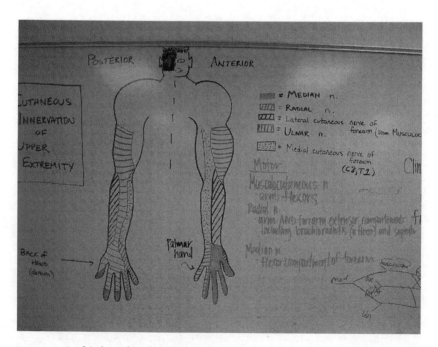

FIGURE 3.4 A whiteboard in the dissection course.

Line Drawings

To read these images, participants move back and forth between the drawing and the accompanying texts, viewing and using the spatial and color-coded relationships depicted. By way of visual-verbal parallelism—parallelism that affords students the opportunity to recognize connections between the drawing and the terminology—whiteboards invite participants to read this juxtaposition as meaningful. By offering a rough sketch instead of a (supposedly) more realistic rendering, schematics emphasize approximate spatial relations coupled with more detailed verbal information on structure–function associations. This flat, 2-D display demonstrates, even provides, an anatomical body alive with biological processes that students then map onto the inert (and dead) cadaveric body. In both courses, whiteboards and the schematics that cover them serve a number of representational and educational functions: (1) they help set the agenda for that day's lab by providing directions and instructions, (2) they reference structures on display, and (3) they provide hints and mnemonics.

First, schematics and the whiteboards in general are agenda-setting tools that TAs and instructors directly reference during prelab talks in both courses. The boards largely display instructions and lists of structures that TAs present in a point-by-point fashion. "Going over the boards" introduces students to the content and aims of that day's lesson. Second, TAs and students understand whiteboard schematics as simplified versions of what they see in other multimodal displays: "Some of the things that are written on the boards are really helpful, because it simplifies things quite a bit as opposed to having every little bit of detail on this and this and this" (Julie, dental student). This simplification serves a number of functions, not the least of which is stress reduction and the elimination of visual clutter. These drawings, created by TAs, are also more authorial and more deliberately conceptual because they represent what the instructors of the course understand as key. Arsenault, Smith, and Beauchamp (2006) observe that scientific drawings seem to "carry the obvious impress of the illustrator's style and selective inclusion" (383). The simplification of figures 3.4, 3.5, and 3.6 offers a less complicated image that stresses specific relations and not others. As such, these schematic displays carry a kind of authorial effect or what Prelli (2006) describes as a "rhetorical selectivity," (12) as students understand this rendering as the result of their teachers' deliberate choices to reveal some features while concealing others. Medical and dental students praised the whiteboards in one particular room because they were, as one student said, "beautiful" and lacked "all that stuff you don't need to see anyway."

The whiteboard in figure 3.5 illustrates the circulatory system of the abdomen, particularly how veins and arteries connect to and facilitate blood flow from the kidneys and adrenal glands. This schematic emphasizes relationships that might not be immediately visible in the cadaver and processes that students might otherwise miss.

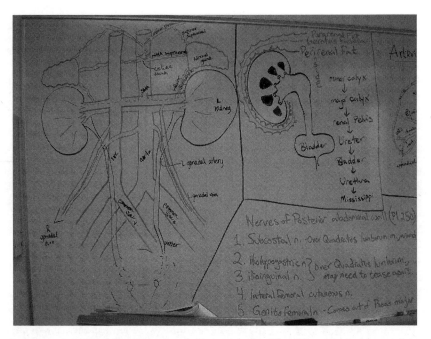

FIGURE 3.5 Whiteboard schematics depicting nerves, blood flow, and organs of the abdomen.

The right side of the whiteboard in figure 3.5 offers a dissected view of a kidney that students will encounter as they make their way through the cadaver. The list at the bottom of the whiteboard offers a quick reminder of the nerves of the posterior abdominal wall, which students will need to recognize and carefully preserve as they dissect. From this material, students gain a conceptual overview of anatomical processes they may not have witnessed in dead tissue. As quick, authorial, and conceptual visual references, schematics recall Bucchi's (2006) view of wall charts in nineteenth-century science education, which offered "a quick and relatively inexpensive" way to communicate descriptions of biological objects and explanations of biological processes (94).

Also, these schematics communicate sometimes by affording a view of the body that students would not otherwise have. Figure 3.6 is a cross-sectional view of the upper torso that "cuts" the body at the shoulders and removes that section (head and all) along the transverse (horizontal) axis. Viewing this drawing is like looking down into the body at the relationships and positions of muscles, arteries, veins, and nerves, all of which are color-coded with a legend on the right side of the drawing.

Students in the dissection course do make such a cut in their cadavers, and this drawing helps them make sense of what they (should) find. However, in the prosection course, with fewer and second-use cadavers, this view into a cadaver

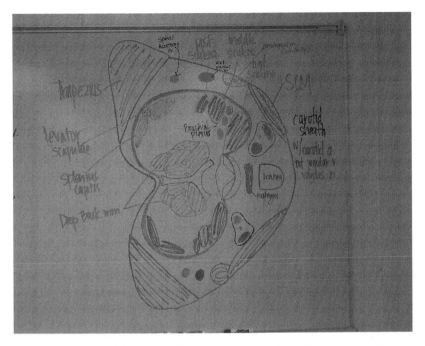

FIGURE 3.6 Whiteboard drawing of the upper torso cut along the transverse plane.

might be impossible. The drawing provides a simple perspective of a complicated dissection that students can reference as they dissect the body (medical and dental students), view the prosections (undergraduates), or learn the information conceptually and not in situ.

TAs and instructors advise students to draw out anatomical structures, to create their own schematics, in order to comprehend, memorize, and later visualize (or call to mind) knowledge of location, function, and shape. Students need to recognize structures across a variety of bodies that do not look exactly alike. Drawing things for themselves overcomes the limitations of textbooks and atlases, whose static images contain limited representations and orientations of structures. These student-generated drawings function as additional and complementary displays that students value more as a process than a final product because rendering the information provides the sought-after opportunity to understand and learn anatomy. In this way, drawing operates as a step toward visualizing, training the mind to memorize structures without having to reference a book or a body. Students also created drawings resembling the heavily labeled schematics on the whiteboard when they felt dissatisfied by either the excessive detail or limited perspective of other displays. Figure 3.7 shows an image of one medical student's drawing of the major circulatory system from heart to groin.

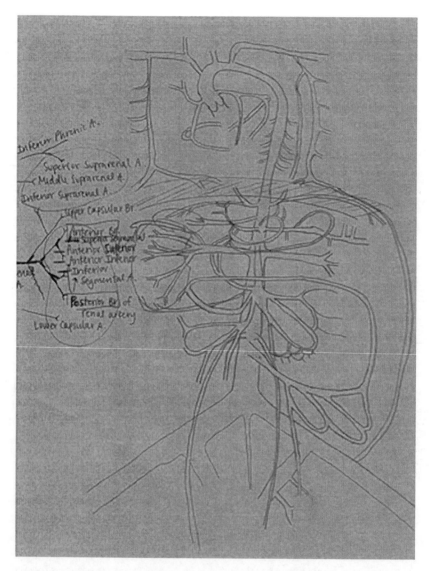

FIGURE 3.7 Student's drawing of arteries and veins of the abdomen.

Patty introduced me to this drawing during our interview, specifically in response to my question about her study habits and the resources she uses, saying, "Oh, and another thing I do when I'm studying for tests is I draw a lot of schematics. So I will—just like what I did last week. It was kind of exciting." At this point in the interview, she moved toward her book bag on the floor beside her, took out the drawing, and began describing it.

I took a piece of paper, and I just started at the heart, and I just started drawing all the branches because like there is no *Netter* plate that does this. I mean, there is, there is a *Body Worlds* person who is all arteries, which I wish I could have in my bedroom, when I'm trying to learn these, you know. But in *Netter*, everything is always broken. And you can see that; you can see this. (Patty, medical student)

Here she passed the drawing to me and continued, "This is two-dimensional, my picture. So this is still not even the best, but this is the kind of stuff that helps me. It is really visual. I am really visual."

Patty's excitement in showing off this drawing, which she calls a schematic, is obvious. She created this image in order to furnish what *Netter* did not, namely a complete rendering of the veins and arteries of the human torso. While this image is not a lifelike rendering, it does provide all the structures with their approximate spatial relations, which she can use to study at home. The ideal display, as she mentions, is actually a plastination of a human body, from von Hagens's *Body Worlds* exhibit, one composed entirely of its arteries. We see her preference for the 3-D anatomy provided by what the exhibit called "Artery Man," for which her line drawing is a humble but useful substitute, one that is still better than *Netter*'s more detailed paintings that have "broken" the arteries. This 2-D drawing, though "not even the best" (her words), affords her a better understanding of and a study aid for memorizing the location and implied function of these structures, thus offering a didactic usefulness not found in perhaps more authentic *Netter* plates. We might even argue that *Netter*'s limitation and her wish to view and understand these structures provided the opportunity for the student-created display.

Verbal Diagrams

Finally, these schematics supply supplemental verbal information such as suggestions and reminders of dissection techniques that TAs and instructors do not always reference directly. Similarly verbal mnemonics are also written out on the board and not always explicitly discussed. Some of these most recited mnemonics offer a simple sentence to encode complex anatomical relationships. For example, the following is an encoded reference to the 12 cranial nerves—the olfactory, optic, oculomotor, trocheal, trigeminal, abducens, facial, vestibulocochlear, glossopharyngeal, vagus, accessory, and hypoglossal nerves: "Oh, Oh, Oh To Touch And Feel Very Good Velvet, AH."[1] In order to remember what type of information these nerves carry (sensory, motor, or both sensory and motor), students add to this the following mnemonic: "Some Say Marry Money, But My Brother Says Big Brains Matter More." Because making sense of mnemonics requires knowledge of anatomical structures and functions, not simply a good memory for the phrases themselves, students use whiteboards to remind them of relationships. One entire whiteboard was taken up by a tabular informational display listing the possible anatomical relationships of the 12 cranial nerves. This elaborate six-column display conveyed information on the name of each cranial nerve, their locations (specifically where each exited the skull), their type (sensory, motor, or both), as well as a sentence-long mnemonic for the name, another mnemonic to remember the type, and a list of either the structures they innervate or the biological functions they make possible (for example, smell, sight, eye muscles).

Naturalistic Images: Didactic Representations II

The second group of 2-D didactic representations is what I call naturalistic images. These include posters, color illustrations, atlas plates, and photographs. These displays demonstrate the human body more directly and realistically by offering a supposedly lifelike representation. Usually taking the form of reproductions of atlas images or purchased wall charts, though instructors and TAs do create their own, naturalistic images serve an expressly authentic and didactic function by providing a more realistic imitation of the body, yet one that is rendered to make visible anatomical relationships and associations. Participants describe these inevitably idealized representations as being like the human body, offering a stronger resemblance that helps students learn and visualize anatomy. The paintings of organs from *Netter*, for example, offer a view of the body more detailed and "real" than the whiteboard drawings. In the prosection and dissection courses, participants used these naturalistic images (1) to assist in identifying, studying, and learning structures; (2) to substitute for cadavers when necessary; and (3) to guide the work of dissection and demonstration. The naturalistic images most prominently used and praised by participants are the idealized anatomical atlas images of *Netter* and the photographs of dissected bodies, both of which afford, even solicit, comparison with the unruly cadaveric body. To interpret and use these images in light of their referent—or to reconcile these pictorial images with the pictorial subjects they index—participants come to appreciate the potential for order that can be imposed on the unruly physical body. This potential for order is primarily what *Netter* and photographic atlases afford.

Atlas Images

Atlases are a visually dominant genre of scientific and medical training guide that offers definitive examples of natural phenomena. Educators, practitioners, and students have used atlases for centuries to train the vision and judgment of both novices and experts. Daston and Galison (2007) describe atlases as "systematic compilations of working objects" (22) that offer "an exemplary form of collective empiricism" that act as a guide for how expert viewers should see and depict phenomena. Atlases train the vision of experts by teaching viewers to recognize the typical, "overlook the incidental," and appreciate the variability of natural phenomena (26). Anatomical science has made liberal use of atlases for centuries, arguably since the publication of Vesalius's *Fabrica* in 1543, though anatomical atlases as we know them—"a compendium of images largely devoid of narrative text"—arguably originate in the nineteenth century (Rosse 1999, 293–94). As a collection of visual representations of anatomical structures, whether paintings, photographs, or computer-generated images, atlases provide idealized and typical renderings of meticulously labeled structures that guide dissections and assist readers in "generating mental pictures" of physical structures in the human body (Rosse 1999, 294).

Netter's *Atlas of Anatomy* is perhaps the most famous atlas of painted images. Figure 3.8 offers a typical *Netter* plate. In this image of the open abdominal cavity, we see primarily the veins of the posterior wall, or the back region of the abdomen. In a field of reddish-pink, the intricate vasculature is identified for easy reference, with the inferior vena cava and abdominal aorta most visible. It is as if the artist, Frank Netter, has masterfully dissected the body of a fit person, one completely devoid of fat and anomalous structures, to demonstrate the major venous system. This image—though idealized and, to some, aesthetically pleasing—is cluttered with labels and corresponding lines that might make it difficult to decipher. Like the whiteboard drawings produced by TAs, this image is a highly authorial and rhetorically selective result of the painter's deliberate choices. For example, a kidney is visible yet unmarked, perhaps as an identifying landmark to aid students as they navigate a more unidealized cadaveric body.

Viewers can see in *Netter* the kidney's relation to these veins, a relationship that becomes a landmark that they can locate as they seek out these same structures in cadavers. The inclusion of this organ and the dissected stubs of major veins underscore the location of these structures and render them worth memorizing, or at least worth noting. It is important to note that this plate is one of the limited views that Patty's drawing (in figure 3.7) is intended to supplement. Atlases, then, participate in the training of anatomical (and later clinical) vision by instructing participants, through repeated exposure, to view the body in a certain light. In the gross lab, this training develops through the networks of images and practices that form anatomical vision.

For most students of my study, Netter's *Atlas* is an indispensable tool. At home or in the lab, students use *Netter* to answer questions and aid dissections. Also, TAs and instructors display *Netter*, or images from it, on the easels and tables, so students can compare them to cadavers and whiteboard schematics. These allow students to cross-reference what they see, or *think* they see, in the actual cadavers:

> So you really need to be able to identify [the structures] on the cadaver, but at least with the pictures you can say, "This is how it should be"

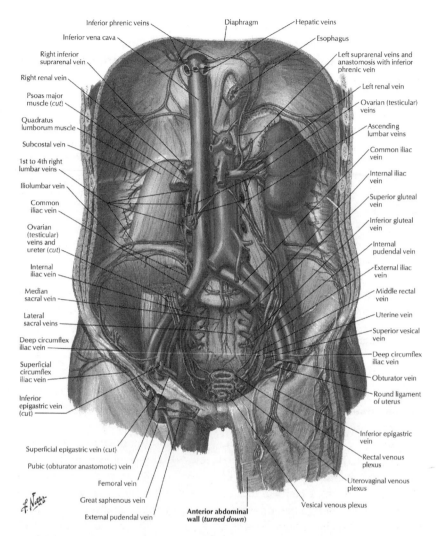

Inferior phrenic veins
Inferior vena cava
Right inferior suprarenal vein
Right renal vein
Psoas major muscle (*cut*)
Quadratus lumborum muscle
Subcostal vein
1st to 4th right lumbar veins
Iliolumbar vein
Common iliac vein
Ovarian (testicular) veins and ureter (*cut*)
Internal iliac vein
Median sacral vein
Lateral sacral veins
Deep circumflex iliac vein
Superficial circumflex iliac vein
Inferior epigastric vein (*cut*)
Superficial epigastric vein (*cut*)
Pubic (obturator anastomotic) vein
Femoral vein
Great saphenous vein
External pudendal vein

Diaphragm
Hepatic veins
Esophagus
Left suprarenal veins and anastomosis with inferior phrenic vein
Left renal vein
Ovarian (testicular) veins
Ascending lumbar veins
Common iliac vein
Internal iliac vein
Superior gluteal vein
Inferior gluteal vein
Internal pudendal vein
External iliac vein
Middle rectal vein
Uterine vein
Superior vesical vein
Deep circumflex iliac vein
Obturator vein
Round ligament of uterus
Inferior epigastric vein
Rectal venous plexus
Uterovaginal venous plexus
Vesical venous plexus

Anterior abdominal wall (*turned down*)

FIGURE 3.8 Plate from Netter's *Atlas* depicting arteries and veins of the abdominal wall. Reprinted with permission from *Netter Images*.

or "how it is generally supposed to be." [You can say] that "this is supposed to be here [and] that "this is supposed to be there." So the pictures sort of give a rough sketch of what it should look like, and then you have to go to the cadavers to say, "O.K. Yes, I now understand how this is put together." (Becky, undergraduate student)

These naturalistic images, according to Becky and others, help students visualize where structures are or should be, and the images do this by displaying the typical shape, location, and landmarks of each structure, information students use to make sense of anatomical relationships.

All the TAs and students I interviewed agreed that naturalistic images, particularly *Netter* plates, regardless of how students use them, should be deployed in conjunction with cadavers:

> Well, I think pictures in books are a lot different than what you actually see, I mean, to look at. You know, you have the cartoon-like pictures in *Netter* that are colored yellow and red and stuff like that, and it's a lot different [than in the body]. With the lab, it is real helpful to see the actual structure [in the body] and see how big it is, what it feels like, and stuff like that. (James, prosection course TA)

The 2-D images cannot communicate adequately scale, depth, or texture, nor can they come close to the authenticity of a human cadaver. The rendering processes of color enhancement and the idealized depictions absent of fat and fascia offer a limited, yet didactic, knowledge of the reality of the body. Though arguably students can and do learn anatomy from these images, they cannot, as the instructor points out, learn what the body is like merely from studying *Netter*-like pictures.

Photographs

Though inferior stand-ins for the body, anatomical photographs of prosections often substitute for cadaveric specimen when students lack immediate access or when there is no cadaveric specimen to demonstrate a particular structure. Whether collected in photographic atlases or provided by the instructors, photographs of cadavers are an excellent example of how naturalistic images function as substitution texts. As Gross (2007) points out, because photographs in biomedical science are believed to "not only show" but also "tell," photographic illustrations are used to "generate scientifically relevant statements" and in the process act as "evidence for these statements" (431). After all, in Sontag's (1977) famous proclamation, "photographs furnish evidence" (5). Several of the medical and dental students admitted to purchasing a photographic atlas, one containing photos of cadaveric prosections, to study at home on days they lacked the time or energy to participate in the afternoon and evening study sessions. But even then, as Emily, a medical student, explained, the photographs were useful only when she already understood the cadaveric body: "I feel that if you're already familiar with the 3-D, looking at the pictures can help." TAs for both courses used photographs and *Netter* plates to prepare for their dissections and teaching sessions. For Richard, a TA in the prosection course, studying with photos prepared him to answer "all the details that [students and other TAs] might ask."

Like drawings, photographs represent structures that the cadavers (on display) at times cannot. A particular structure in question might not be present in the cadavers if, for example, the structure does not exist in a typical state in a dead body or if someone has already dissected that structure away. As Belinda, a TA in the prosection course, put it, "[pictures] also just reinforce things—things that we can't dissect, things that we can't show." One key photograph in the undergraduate lab depicts a close up of an open, brainless cranium, spotlighting a network of arteries called "the Circle of Willis" (figure 3.9).

TAs slip this printed image safely inside a plastic cover and place it on an easel close to a cadaver that demonstrates the structures inside an open, empty cranium. To render the Circle of Willis more visible in a cadaver and thus to increase its didactic value, someone has to inject the arteries with red silicon, which dyes and inflates them, making them appear more like they would in a living body. Though the instructors can and have created this more "natural" example, they can only do so if there is a cadaver with a suitable (usable) Circle of Willis. Often, a photo is used instead, illustrating what is absent in the body next to it. Photographs of anatomy, whether created by instructors or atlas editors, carry an authorial stamp not unlike a drawing or painting, in that the image is a result of someone's decision that this structure, viewed in this way, is a worthy object for anatomical study. These photographs, in the words of Berger (1980), "bear witness to a human choice" made by instructors and TAs who select and authorize these views and perspectives of the body (293).

FIGURE 3.9 Photograph of a cranium with a visible Circle of Willis.

Photographs help somewhat to satisfy the authentic and didactic requirements of anatomical illustrations because they provide what is taken to be docu-
mentary evidence of real human anatomy. Arsenault, Smith, and Beauchamp
(2006) understand photographic viewing as a mimetic exercise made possible
by the isometric (or proportional) relations between photos of objects and those
actual objects. Sekula (1984) describes this relation between a photograph's pictorial image and its pictorial subject as a metonymic process of substitution (10).
Through pattern recognition and a kind of "spatial mapping," looking at photographs imitates looking at objects in the world (Arsenault, Smith, and Beauchamp 2006, 383). Because the representational content (or pictorial image) of
photographs can be mapped onto the body, photos aid students and TAs in what
I identify as the third purpose of these naturalistic images: assisting in dissecting
and, thus, "revealing" anatomy. Repeatedly, TAs and students praise the assistance
naturalistic images (primarily *Netter*) provided as they "made their way through
the body" (an expression common among participants). The following interview
excerpt describes photographs and *Netter* plates as dissection tools:

> Oh, I think *Netter* is always helpful for when you're trying to dissect because
> you don't know what you're looking for, where you're going, and otherwise
> you're just going to annihilate your body. So *Netter* and the photo atlas
> thing is good. (Marianna, dental student)

These naturalistic images guide dissection by providing an image of idealized
anatomy that functions as a set of visual instructions, a roadmap that helps dissectors know what to look for and where it should be found. Otherwise, dissecting by
trial-and-error or by way of verbal descriptions from whiteboards or *Grant's Dissector* could result in a destruction of the "usefulness" of the body. If a cadaver loses
its ability to represent anatomy (that is, to become the anatomical body), then that
cadaver loses its didactic value. Thus TAs and students map the idealized images
onto the unruly cadaveric body, and then they map this unruly body (with all its
restrictions) onto either *Netter* or photographs. Participants use these images as
visualization tools to create an image of the body that makes dissection possible.

> Well, I think the actual cadaver is better for learning than *Netter* is. I think
> *Netter* helps me dissect, but the body is better for learning because there are so
> many anomalies [in the body]. And *Netter* is, you know, pretty [laughs]. They
> call it "netterized," like, you know, because it is perfect almost, those images.
> (Marianna, dental student)

Near perfect, "netterized" images help students and TAs transform the body
into a 3-D display of anatomy, a "cleaned up" (as many participants described
it) object participants use to learn. But the isometric compatibility of the images

and the body, or their supposed correspondence, is limited by the idealization of these static, flat, 2-D renderings. After all, the diversity of anomalies, anomalies that *Netter* or a photograph of one body cannot represent, is part of anatomical knowledge. Visualizations offered by the images are only an initial component of learning; learning must ultimately come back to the cadaveric body:

> So the cadaver that I learn from the *Netter* helps me dissect. And if I'm trying to study, and I cannot be in the lab, it is nice to look in *Netter* and the photos just to remind myself of where it was, so that I can then picture it on the cadaver, picture where it was on the cadaver. (Marianna, dental student)

I am struck by her phrase "the cadaver that I learn from the *Netter*." *Netter*, photographs, and other naturalistic images offer viewers not only an image of the body but *a certain type of body*, one of important but limited usefulness. Again, these images are substitutions when participants are without a cadaver, but only if they keep in mind what one student described as the theoretical nature of these images. The naturalistic displays demonstrate anatomy by seeming to resemble bodies (*Netter*) and sometimes depicting bodies (photographs), but they do not provide the "reality" of anatomy, which can only be experienced in the physical body. Students and TAs often deploy naturalistic images in the absence of the real (original) object in order to "picture" or visualize (recall) the cadaveric body. When used to identify features of the physical body, these displays solicit comparison through a presumed correspondence to the referent. Making sense of the original object in light of the representation means making sense of or performing some task with the real object (the body) by way of this display (*Netter*). The representation, then, affords a view of the original, or affords actions involving the original, that participants cannot always get from or accomplish with the original itself.

Nonhuman Corporeal Objects: Didactic Representations III

Plastic Body Models

The plastic models, used mostly in the prosection course, are obviously not human bodies but represent human anatomy by way of their realistic, 3-D likeness. As Van Dijck (2005) reminds us, "body models" have been used as pedagogical aids in medical education "since the early modern periods." Their common use is due in part to the didactic value they afford, namely that features "can be disproportionately accentuated" to communicate "particular anatomical insights" (44). The plastic models shown in figure 3.10 display vivid color enhancement that, though not resembling the living or even dead body, does offer a clear view of structural relationships and distinct (perhaps exaggerated) demarcations between structures.

FIGURE 3.10 Plastic models of structures from the abdominal region.

The at-a-glance obviousness of these plastic models makes them excellent tools for communicating the interworking of veins, arteries, and ducts in the hemisected kidney (in the box in the middle of the table in figure 3.10). However, their media make them less ideal than authentic specimen. Because they are plastic, rigid, smooth, unmalleable, and idealized, students cannot learn the "organic texture" of the human body by handling what are essentially anatomical dolls (Van Dijck 2005, 44). In a discussion of the importance of cadavers, one male medical student, Andy, explains the limitations of plastic models:

> I think a mannequin factory would probably make them all the same, whereas the human obviously can have variation. So I think that is part of the important thing about using cadavers: that you can understand that we are all different. By understanding the differences, you can understand how people anatomically are different.

The exaggerated idealization and static rigidity of these mannequin-like objects prevent a student from fully understanding the reality of human anatomical variation. Understanding this variation across a range of cadavers is important to these future doctors and dentists, as they must deal with the singular, yet multiple, human body. Because students are examined on actual cadavers and not idealized plastic bodies, they also must recognize and appreciate anatomical variation in the here and now. (As I discuss in the next chapter, the authenticity of human bodies signifies by way of haptic or touch-based experiences of texture, smoothness, depth, and feel.)

Nonhuman Corporeal Images: Authentic Representations

Participants understand what I call authentic representations not as substitutes for cadaveric specimens but as a specialized lens that makes them visible in a new way. For those not trained in anatomy, radiographic or x-ray representations are no doubt just that—representations of bodies. To participants engaged in anatomical practices, these representations of the anatomical body were seen as more authentic. Participants understood these technological mediations of actual cadavers—the radiographic, CT, and PET images—as closer to real bodies than the naturalistic images from *Netter*. This group of nonhuman corporeal images illustrates the power of socialization for understanding how visual displays of science facilitate meaning, action, and perception formation. In the labs I observed, x-rays of cadavers were the most common nonhuman corporeal image. By seizing the unique representational affordances of x-rays, students learn to recognize anatomy in these authentic representations. They do this by comparing the pictorial image of these medical displays with that provided by schematics, naturalistic images, and cadavers.

Radiographic Images

Radiographic, or x-ray, images and other forms of medical imaging, like PET scans and CT images, afford a view beneath uncut, undissected skin, a macroscopic look at gross anatomy untouched by human hands. In the case of x-rays, participants deemed the uniqueness of this affordance more significant than the formal properties of the image—namely that it is a specialized, 2-D photograph. Figure 3.11 illustrates a typical example of the lab's radiographic displays.

A TA or instructor has tagged the x-ray on the left with arrows that call attention to specific anatomical structures. Historically speaking, radiographic technology, according to Van Dijck (2005), creates "an illusion of unmediated, objective reality," not unlike photography (98). Pasveer (2006) similarly asserts that x-rays are often "believed to produce objective renderings of existing" phenomena normally invisible to ordinary practices of looking (42). The reality of radiographic technologies, however, is that they produce highly mediated images that we understand only through specific training and knowledge. Participants read an x-ray as an image of the body only after they have learned the requisite anatomy to make sense of what are essentially shadows on film. What is revealed in the image is the instantiation of anatomical knowledge and the rendering or conflation of the physical body as the anatomical body, a conflation that is a consequence of anatomical vision. For students and TAs, these mediating and rendering processes were only significant as initial obstacles to be overcome through

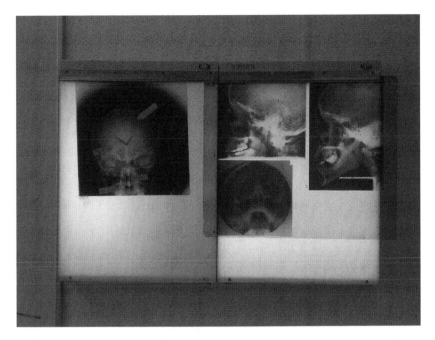

FIGURE 3.11 Radiographs of skulls on a light box.

training. In other words, several students described initial difficulty in reading x-rays because they never had to make meaning from a radiographic image. Once they understood how to read this mediated display, they soon discovered that, as one male student put it, they had to know "the anatomical names and structures or else it was just a crazy x-ray." But they learn the necessary anatomical knowledge and how to recognize that knowledge in the x-ray by learning to respond to the opportunities x-rays make possible. As Hutchins (2010) describes it, students learn to understand the image an x-ray provides through skillfully enacting meaning. Specifically, x-rays afford a peek at the interior of the uncut body, offering anatomical knowledge of "real" anatomy, even though they are still little more than 2-D images.

Human Corporeal Images: Authentic Presentations I

Authentic presentations of human corporeal displays arguably calls into question any wish to conceive of images as representations and objects as presentations. These images, primarily microscopy, are more "than just pictures" (one female TA's phrase) because they contain real human material. The course instructors prepare the microscope slides by cutting extremely thin sections from human organs or tissue. As the course instructor explained, the tissues are preserved with a fixative (formaldehyde, gluteraldehyde, sometimes even ethanol), and then stained with the appropriate chemical, based on the specific structure, compound, or molecule to be presented. Microscopy affords an opportunity to demonstrate "real" or, as another TA put it, "direct anatomy." These displays make visible microscopic anatomy, one of the avenues by which anatomy is taught and studied. (The three other ways of studying anatomy include radiographic anatomy, embryological or developmental anatomy, and, of course, macroscopic or gross anatomy—anatomy visible to the naked eye.)

Microscopy

TAs and instructors exhibit histological images either by way of the microscopes themselves or through (8.5 × 11 inch) color prints of microscope slides. The microscope images are understood by Cartwright (1995) as technological media- tions of the interiority of the body that inevitably present an image of the body as fragmented, "excised," and "stripped of its corporeality" (83). Accordingly, microscopy offers an image of miniscule interiority that is not only separated from the rest of the body, but that also erases the cultural constructions of per- sonhood that we, as humans, both embrace and resist. These are, I would argue, possible ideological effects not necessarily rooted in the purpose of the images or the circumstances of their creation but in the affordances the pictorial vehicle offers the trained participant. Microscopy, historically speaking, is a product of what Daston and Galison (2007) term "mechanical objectivity," or an impulse in scientific illustration practices to remove the subjectivity and the fallibly of the human eye (43). By rendering visible otherwise invisible human tissue, micros- copy affords a view of isolated cells, a glimpse of real anatomy (due to the human tissue on the slides), but they afford this glimpse because of the technological mediation of the apparatus itself.

As students maneuver the labs, the microscopes on the table afford looking into; they are look-into-able, so to speak. In using them, students are not only looking at pictures of anatomy (pictorial images); they are looking into a viewing apparatus that enhances their perception (an affordance) and offers examples of anatomy (pictorial subjects). For the TAs in both labs, the affordances of micro- scope images were categorically more important than their textuality as special- ized photographs. One female TA explained that a photo of gross anatomy, like those in the photographic atlas, was not as significant as microscopy because the former was a poor substitute for the cadaver, whereas the latter was the authentic way to see those structures. Microscopy was not a substitute for human tissue; it was a view of the tissue itself. They, of course, admitted the complications of microscopy: it is generated by technology; it is something they learn to use; it is not always well suited for some students (those with glasses or even long eye- lashes, one TA joked). Often instructors and TAs display other images next to the microscopes, as figure 3.12 shows. These additional images, such as photographic images of microscopic tissue, are placed beside the microscopes to offer students a comparative perspective. If they cannot easily make out the picture the microscope

FIGURE 3.12 Microscopes displayed with microscopy photographs.

slide offers, they may look at the printed image beside it. Together these different types of visual displays provide different vantage points for the same anatomical phenomena.

At times, the technological mediation of microscopy interfered with the rendering practices needed to create "representative" anatomical specimen. During prep labs for the prosection course, I witnessed TAs struggle to overcome the resistance microscopy at times provided. Here is a field-note excerpt depicting a group of TAs setting up the microscope slides for an undergraduate practical exam:

> Back at the microscopes to my right, one of the female TAs is struggling to find representative examples of a striated muscle.
>
> First TA: What is that?
> Second TA: A smooth muscle?
> First TA: Oh, man. [She is visibly disappointed.]
> Second TA: [No reply]
> First TA: Can you see the striations?
>
> A third and older female TA has walked over and is talking to the second TA, who has already finished setting up her microscopes. The third female praises as "excellent" the slides the second female has set out.

First TA: Look at this.
Third TA: Is it smooth?
First TA: Ahhh. [She searches for another slide.] These are so unrecognizable for cardiac.
Third TA: Use the others.
First TA: I have tried them all. (Field note, prosection course prep lab)

The first TA struggles to find a recognizable and "realistic" example of striated muscle. Here her role as TA and her knowledge of anatomy conflict with the histological slides available to her. While she might criticize the condition of the teaching material she must use, we might view this as proof of her advanced anatomical vision, which allows her to recognize the limitations of these slides, namely that they are unrecognizable as examples of what they actually are. Lynch (1985) would no doubt consider this an example of the possible recalcitrance of the docile object of laboratory work (44). At the risk of ascribing agency to objects, the slides resist the rendering practices that make them useful for the task. Thus, the first TA repeatedly enlists the assistance of her colleagues by asking them to look at the slides and lend their skills in confirming her suspicions. Even for these more advanced participants (TAs), viewing histological anatomy can be complicated by the technological mediation that makes microscopy significant.

Human Corporeal Objects: Authentic Presentations II

The participants of my study viewed the human corporeal objects, more than any other type of visual display, as the ultimate learning tool and the definitive visual text, offering the "reality check," as one male dental student described it, needed to learn and teach anatomy. These objects can be divided into four types: (1) plastinated human tissue, (2) excised human tissue, (3) human cadavers (presented whole or in excised parts), and (4) living human bodies. Like the 2-D images, these 3-D objects, whose media range from plastic (the models) to human tissue (the cadavers) to something in between (the plastinations), are products of the same rendering processes of enaction (through recognition) that transform the physical body into the anatomical body. The visual and tactile features of these objects, specifically the bodies (whether excised, dissected, or living and intact), invite participants to look and touch. By working with the affordances of these objects, perhaps more than any other, participants learn to interpret all bodies as potential anatomical specimen—thus enacting the anatomical body by way of the physical human one.

Plastinations

The recent biotechnological development of plastination has nearly reconciled the often-competing values of didacticism and authenticity. This process, created by the German anatomist Gunter von Hagens, allows anatomists to render human tissue into a more durable, transportable, and preservable state. Through a process that halts tissue decay, plastination replaces the body's liquids with acetone, which is then replaced with a polymer solution. This process of replacement, which von Hagens terms "forced impregnations," infuses every cell of the body with the liquid polymer (von Hagens, Tiedemann, and Kriz 1987, 417). The human body or excised tissue, positioned in a particular pose, hardens by using either light, heat, or gas. The result is an object that is technically both human and nonhuman in that each cell of this otherwise biological body is infused with plastic. This process, according to Van Dijck (2005), moves "beyond the body-or-model dilemma" because plastinations are "real and modifiable," proving that the didactic and authentic usefulness of anatomical teaching objects is not an either-or proposition (49). Students can see human variation and actual anatomy in the plastic-like model that still retains some of the natural texture and feel of (nonplastinated) humans. To use Andy's metaphor, plastinations are not mannequins though participants can handle them as if they were, without fear of damaging the now-durable specimen.

Set out on a separate table away from the cadavers, the plastinations were primarily displayed in the prosection course. Signs reminded students to remove their gloves before touching surrounded plastinations, which were mostly internal organs and not whole-body plastinations like those in *Body Worlds*. Because these objects were as much plastic as human, handling them with gloves greasy from cadaveric tissue could possibly cause mold or fungal growth and make them unusable in the future. Standing around these tables, a few undergraduates explained to me "the weirdness" (as one female student put it) of touching "bodies that aren't bodies."

Excised Human Tissue

The most common form of human body model was not the plastic or plastinated models but the excised human bones and organs displayed on cadaver tanks. These corporeal objects serve as the referent which all schematics, naturalistic images, and plastic corporeal objects are said to represent. The excised structures, for example, bones of the skull (figure 3.13), are authentic human structures that TAs or instructors have dissected and removed from the body.

Participants can touch (with gloves) and view from different angles these more manipulatable objects. By doing so, they gain a better understanding of structures and their relationships to neighboring landmarks and systems. Students could take individual cranial bones, such as those depicted in figure 3.13, and piece

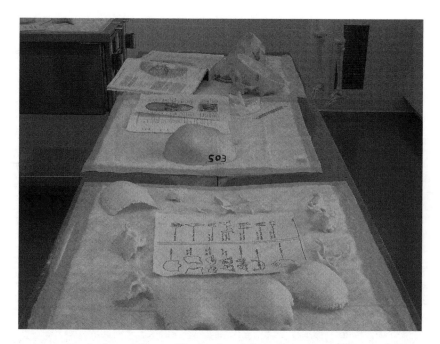

FIGURE 3.13 Excised cranial bones displayed with Netter's *Atlas* and drawings.

them back together to reconstruct a complete human skull. Another lab session displayed excised vertebrae that students could stack to physically reconstruct and feel the architecture of the backbone. As in the figure above, TAs and instructors often display these excised specimens alongside naturalistic images (like *Netter*) and drawings, to provide an additional view of the object as well as labels and legend. Obviously these 3-D objects—both excised tissues and plastinations—afford students an embodied and tactile knowledge of human anatomy by adding the experiences of touch that 2-D images cannot. I witnessed students in the prosection course holding and touching excised portions of the digestive tract, from stomach all the way to anus, which was displayed during the digestive system lab. One female student explained her surprised comprehension of the various sphincters that make up the tract. Though she knew they existed from textbook descriptions, the images in *Netter*, and the course lecture, she did not fully "get it" (her words) until she could touch for herself the pyloric sphincter muscle that is the gateway to the stomach.

Cadavers

The human cadaver is the corporeal object that marks the gross lab as a unique classroom and an important medical experience. The cadavers are, according to participants of my study, "real anatomy." Students, instructors, and TAs for both courses describe cadavers as "indispensable," "priceless," "the most valuable thing," and "the reason for the course." More than schematics and naturalistic images, cadavers, specifically their use in dissection and demonstration, are the rationale behind gross anatomy labs—the teaching and learning of 3-D anatomy. A male medical student, Samuel, best expresses the philosophy behind cadaveric anatomy and the use of cadavers as multimodal displays, which I call the "reality effect":

> Um, I think that [the cadavers] are fantastic because everything else is just theory. I mean the notes are just theory. The *Netter* plates are theory. It's all about "this is what it is most of the time." And then you get to the body and, you know, things change. And arteries, and that is probably the biggest thing, they can go anywhere. Like our body [our cadaver] had three pudendal arteries coming out all different directions.

cadaver also theory

He continues,

> So I think that is what they are the greatest use for, realizing that this is not what we learn in class, and what we see in the book is not what it actually is in the human body. It is just a guide for where to start.

First, Samuel describes the cadaveric body as both the referent (or pictorial subject) to all other displays and the reality of anatomy. In a sense, schematics and naturalistic displays are merely conceptual or theoretical substitutes that reduce a complex 3-D object into a 2-D representation. The cadaver is the reality on which those concepts are founded. Rather than view cadavers as instantiations of anatomical knowledge made material, Samuel suggests that the cadaver offers the truth that is the body. It is not that Samuel and his fellow students use anatomical knowledge to interpret the human body as an anatomical object; instead, the human body offers lab participants the reality of anatomical knowledge. The "reality effect" is this idea that the anatomical body is the physical body, and that the physical body is anatomy. Cadavers afford this reality effect through the lab's embodied and rhetorical practices. This formation of the body's twofold singularity recalls Merleau-Ponty's (1968) assertion that we are a "phenomenal body" and "objective body," a body sentient and able to be sensed, perceiving and able to be perceived (136). Though participants experience this reality effect in relation to a number of multimodal displays, human cadavers seem to solidify this configuration.

Samuel, then, backs up his theory-versus-reality discussion with an example of how a cadaver's arterial system can manifest wide variation in location and number. His team's cadaver instructs them on the complexity of anatomical variation through the physical reality of its body: the cadaver had three pudendal arteries instead of two. By "coming out in all directions," Samuel vividly expresses the way this reality effect is best communicated perhaps in unruly cadaveric bodies and not idealized, netterized naturalistic images and plastic models. By this logic, the reality of the body is often chaotic. To learn anatomy, as Samuel asserts, students have to understand the difference between idealized images and real, unidealized bodies (between pictorial images and pictorial subjects). Schematics, naturalistic images, and nonhuman corporeal objects afford only conceptual information that participants apply to the body's reality. The best way to apply these idealizations to reality is through (1) dissection, which transforms the body into a multimodal display by controlling the chaos; and (2) demonstrations and observations, which read the physical body as anatomical by mapping anatomical knowledge onto that body. Dissecting, demonstrating, and observing enact the anatomical body out of the flesh of the physical body. This transformation is possible because participants learn to perceive and interpret the cadaver's affordances.

Living Bodies

Cadavers are not the only human bodies of the lab to serve as multimodal displays. The bodies of everyone in that space potentially serve as didactic and authentic presentations and embodiments of anatomical knowledge. Whenever a participant uses a human body to demonstrate anatomy, that particular body (living or dead) instantiates and materializes the anatomical body. Often, this use of the living body as a demonstrational tool is not merely an inadvertent consequence of laboratory practices but a deliberate enactment (or performance) of anatomical knowledge. Students and TAs frequently turn to their own bodies to illustrate structures or structure–function relationships not recognizable or easily demonstrable on cadavers. For example:

> When we were doing the arm muscles it was really helpful to be able to look at someone who was really ripped [muscular] and had very defined muscles. So that I could be like, "do this" [ask the person to flex]. Then I could see it on the cadaver, and I could see where it is on the real person. And it helps to make that, you know, just to solidify where it is. (Julie, dental student)

Julie explains how participants can more easily understand the physiology of particular muscles on living humans because cadavers cannot demonstrate motion, and because the cadavers, often the bodies of elderly people, have suffered atrophy before death. Here Julia tells a familiar story witnessed by anyone observing the gross lab: students move their attention away from cadaveric bodies and onto living ones that provide solicitations to action that the dead cannot.

Living bodies also become corporeal objects of display when students use them to demonstrate common body positions they must remember. The most popular anatomical performance is perhaps the gestures used to demonstrate the function of the legs' extensor and flexor muscles. To express the motions made possible by the flexor muscles, participants stand on the heels of their feet and then flex their toes up toward their knees, bearing their body weight on their heels. For the extensors, they raise and lower their bodies on the balls of their feet, as though they were wearing invisible high heels. These simple bodily gestures convey anatomical knowledge of muscle function and serve as visually embodied mnemonics students incorporate into their learning. In fact, during two different exams (one practice exam and one actual exam), I observed three students mimic the extension and flexion motion as they stood near two particular tables. In the practice exam, a female TA reminded students not to gesture with their bodies because it might suggest an answer to their classmates. (Perhaps an incorrect answer; one student enacted these mnemonics at the wrong table.)

Such anatomical performances inevitably encourage participants to personalize the anatomical body, thus mapping the anatomical body onto their own. Any demonstration activity can become an opportunity for embodied internalizations of anatomy. During one demonstration in the prosection course, a male TA, seeing students confused about the correct orientation of an excised scapula, held the bone up to his own body. By positioning this scapula atop his own scapula, he displayed the bone's location in the body. During a prelab talk in that same course, the instructor used his hands to demonstrate what is often termed the "anatomical snuff box." This deepened, triangular spot on the back (or dorsum) of the hand is at the level of the carpal bones where the thumb meets the wrist. As the instructor demonstrated this anatomical location, he asked the class to feel their own hands (and "snuff boxes") to understand the structure. In this self-demonstration, the instructor offers a visually embodied mnemonic (the gesture) and a discursive mnemonic (the phrase "snuff box") that students not only mimicked but later discussed among themselves—the reference to snuff became for many students a memorable part of the demonstration in that several of them confessed to complete ignorance about the tobacco-based product.

Performances of the anatomical body such as these can be deliberate pantomimes that demonstrate aspects of the anatomical body by making it visible on the living human body. But these performances can also originate with the human body's capacities for action. Early in the semester, I watched a body-buddy team leaning over its cadaver, as one male student asked his teammates what they should see in the axilla region. A female student replied, "The subscapular artery." The same male student then asked her to explain: "How does that work?" The other female student at the table, near the head of the cadaver, explained how the subscapular artery is oriented in relation to the serratus anterior muscles. To show this, she used her two hands placed in front of her. Through gesture, she demonstrated how the artery would run around the structures. As she struggled to figure out how exactly to perform her manual demonstration, the other male student completed her performance. He did this by pointing to her hands and not the body, using her hands as part of his explanation. Here the living body is used, through gestural performance, to call attention to specific details of the cadaveric body and specific steps involved in dissecting it. These students not only inscribed anatomical knowledge on the cadaver but also on their own bodies, incorporating this now embodied knowledge into their demonstration. And they did this because the cadavers made visual and tactile recognition of this particular structure difficult. Thus, they used their own bodies to enact recognition.

These embodied performances, which McNeill (2005) might define as "gesticulation" (12), occur as a consequence of dealing with the communication difficulties of the work, as Sauer (2003) discusses in her study of gestural communication among miners and their families. As the scene illustrates, some knowledge (some ways of viewing the world) are more easily communicated through embodied movements, particularly when students have wet, gloved hands. Moments like this are common and illustrate the ways the anatomical body is enacted through participants' bodily engagements with the affordances of the objects that surround them. And this enactment of meaning happens in part because the living human body, as a mobile, tactile, 3-D object, can easily be positioned to demonstrate anatomy. In this space, human bodies provide opportunities to perform and become the anatomical body because of these affordances. Through trained vision, participants learn that anatomy is a system that can be mapped onto and discovered in the body. Students and even TAs learn this through their embodied engagements with the multimodal displays, whose representational nature participants enact by way of their own bodily capacities for meaning making.

Conclusion

In the gross lab, these visually and haptically oriented multimodal displays—schematics, naturalistic images, corporeal images, and corporeal objects—differently instantiate, mediate, and materialize the anatomical body. These images and objects are deployed based on their properties: (1) the display's mediation, or its representational or pictorial content, which often determines its authenticity or realness; (2) the object's materiality and media, along with the assumptions we make about those media; and (3) the object's dimensionality and level of interaction or, specifically, its object-ness, which can convey anatomical knowledge as well as the opportunities for action that allow participants to make sense of anatomical knowledge and project it onto physical bodies. Though all displays are important pedagogical tools, participants understood some objects (cadavers, x-rays, and microscopic images) as affording knowledge of real anatomy, whereas they perceived others (photographs, atlas images, and plastic models) as mere stand-ins for the real thing. Participants understood 3-D human corporeal objects, especially living and dead bodies, as physical, corporeal, and manipulable presentations of authentic anatomy. The object-ness, the materiality, of these multimodal displays mattered a great deal. The naturalistic images are limited by their flat two-dimensionality, just as the human corporeal objects (plastinations and cadavers) are lauded for their three-dimensionality and haptic authenticity as human tissue.

To close, I offer a narrative illustration. Late one afternoon, during an open lab period in the dissection course, the student Ted jotted down notes from the whiteboards. I asked him how things were going. "Slim pickings," he said, pointing to the information spread across the boards in blue and red marker. He went on to explain that there was not that much "that grabs me" today. He clarified that nothing on the boards really warranted copying because he either knew it already or had it in another, more helpful form. When I ask him what he was writing, he turned the notebook so I could see. On the page was a drawing not transcribed from the board but created by him from the board's information. He explained that "this will help me get it, teach it to myself later." He told me that if he could draw it and if he could "make it go where it needs to," then he would know he "really knew it."

This example of a student at the whiteboard highlights the ways we enact an object's affordances through our intentional interactions with them in settings saturated with social and cultural meaning. Rather than copy the information on display, Ted transforms that information into a visual display of his own that is intended to help him understand and learn anatomical relationships. He perceives a particular affordance in his social environment, one offered by the whiteboard and the information on it. But more than that, he recognizes in them an opportunity to render this information into a different form. This drawing, a schematic representation, will become a text, an instrument, and a perceptual tool he can use to make sense of the body, again because of some affordance involved in the process of making the drawing and the product of the drawing itself. This drawing is a representation of the anatomical body, a body that due to his socialization as an anatomy student is merging more and more with the physical body of lived experience.

As students engage in and learn anatomy by way of these practices, they come to see visual objects as affording opportunities for action. What they find when they look at these visual displays and objects is a series of bodies and not-bodies, both of which become anatomical bodies, both of which are constituted by the object's mediation, dimensional, materiality, and level of interaction, all of which participants perceive as providing embodied opportunities. These displays then facilitate, within the perception of the trained viewer, a way of seeing that mutually articulates the physical body and the anatomical body as one and the same. By way of this trained vision, the physical human body is not inscribed with anatomical knowledge, thus becoming the anatomical body; instead, the physical body *is* the anatomical body. The anatomical body is enacted through participants' embodied engagements with the affordances of the multimodal objects that surround them, affordances that are neither solely in the object nor solely in the user but that emerge through the interaction of the two. These multimodal displays exert a persuasive and ontological force in that they are used to teach and learn anatomical knowledge and to enact this knowledge in and on physical bodies in a way that mutually articulates and enacts both simultaneously. This trained perspective that enacts the anatomical body is a consequence of the embodied rhetorical actions of the lab.

In any technical domain, the multimodal displays that instantiate and facilitate expertise operate as rhetorical objects that induce in participants actions and attitudes, ways of seeing, moving, and being in the world. Yet research into multimodality frequently focuses on the supposed "gains and losses" involved in translating information from one mode or media to another (Kress 2005, 5). Though offering helpful insights, this line of inquiry inevitably conceives of information as easily contained in a text or object that we access by using the object's pregiven affordances. Contrary to this, I want to suggest that we enact an object's information through our physical, embodied interactions with that object's affordances, which are opportunities for action that emerge from the mutual contact between the object-ness of the object and our bodily capacities for perception, movement, interpretation, and meaning making. Affordances of displays and objects are not as much perceived as they are enacted or made through our purpose-driven interactions with them. Affordances are enacted through the body-object-environment assemblage.

This insight has real consequences for the way we conceive of technical communication, particularly the production and deployment of multimodal displays. For example, visual and verbal schematics—from technical diagrams to blueprints—are everywhere in TPC. We accomplish a great deal of both complex and mundane work, from programming a computer to learning a simple software application, with these displays we often take for granted. Though simplistic displays might not require a great deal of embodied activity—except the movements required for reading signage or using a diagram in a manual—inevitably they invite specific actions from the viewer engaged in meaning-making tasks by seeming to provide specific opportunities for that viewer to engage in those tasks. The schematic rendering a drawing provides easily emphasizes certain features and relationships. When a user, whether an engineer, student, or professional writer, engages in a task using this multimodal display, the schematic encourages that user to notice and act upon features and affordance that are coproduced by the user and the display. The user and the display interact in a socially and culturally saturated context. An exploded view diagram of a gear pump, then, is not just an illustration communicating information about an object; it is a display that provides possibilities for action when we engage it in authorized ways, in situated contexts of use. Every day, technical communicators are engaged in these interpretative and interactional tasks. When we label or verbally describe a naturalistic display, we solicit from future users embodied actions, interpretative practices, and even conceptualizations of that display that those users can only discover by engaging with those informational objects in an embodied, skillful, and intentional way. More than just communicating about technology, technical communicators use multimodal displays to educate and even to socialize users to observe, interpret, and act upon the objects that we create. And through the interface of bodies and objects, meaning is enacted.

4

HANDS-ON VISUALS

Embodied Observation and Rhetorical Verification

In discussing the significance of autopsy, Saunders (2008) traces the rhetorical roots of this medical practice. Derived from the Greek for "to see with one's own eyes," autopsy suggests the firsthand observation of bodies by medical professionals, a supposedly direct perception that calls to mind the "testimony of the eyewitness" (16). Autopsy, Saunders contends, "connotes a problematic of persuasion and reception," a "representational and rhetorical economy" that privileges certain ways of seeing and bearing witness to what is seen (17). During an autopsy, a trained professional draws upon her expertise and skills in medical observation to make sense of the body in front of her. While gross anatomy students do not perform forensic autopsies to determine cause of death, their postmortem examinations of cadavers do require similar skills of observation and deliberation. These processes of debate and confirmation recall McKeon's (1987) discussion of the subtle connections among demonstration, verification, and justification, an interrelation that, I argue, is particularly salient in medical education. Through their firsthand observations of bodies, students are required to provide "an exhibition, portrayal, or presentation" of "the noteworthy characteristics" of the body on display. They perform these demonstrations, either for their classmates or themselves, by communicating their "reasons for crediting conclusions from experience or knowledge" (39). These entangled practices of demonstrating and verifying anatomical evidence require both observational and rhetorical skills.

In learning to interact with and make sense of multimodal displays, students must constantly verify and even assess their own learning; they must confirm for themselves what they do and do not know. For example, one undergraduate described for me his morning routine of note taking, drawing, and reviewing *Netter*:

I will physically show myself if I can do this and this and this. "What are the names of these?" I ask myself. So this is how I do it. This is how I study. But I am a very hands-on, a very visual person. I study visually, and I teach myself visually as well. (Rodney, undergraduate student)

Rodney describes using the course text (*Netter*) to create more texts (drawings and notes) that he uses to demonstrate and even persuade himself that he has learned. By connecting these steps to a type of visual learning that is "hands-on," he emphasizes the importance of visualizing anatomy and its dependence on bodily actions, specifically ones that are kinesthetic (movement-based) and haptic (touch-based) ways of knowing.

At the heart of Rodney's self-assessment rests a kind of self-interrogation. He uses these displays and objects to convince himself he is not just memorizing names or guessing at structures, but that he understands anatomy in all its complexity and variation. This process of confirmation is not unlike Nienkamp's (2009) idea of "cultivated internal rhetoric," a "learned and deliberately cultivated practice of internal self-persuasion" (19) by which we talk ourselves into certain "actions, attitudes, or beliefs" (18). For Nienkamp (2001), internal rhetorics develop from the rhetorical interaction of "internalized social languages" that constitute the "rhetorical self" (127). In the gross lab, these acts of self-persuasion, which contribute to trained vision, require participants to internalize language, in this case, anatomical discourse. In addition, students must internalize the visual and physical features, so to speak, of anatomical structures. They do this by learning the look and feel, the descriptions and relationships, of the cadaveric body—the lab's most complex multimodal display.

Building on my previous discussion of how participants make sense of displays by recognizing and seizing their enacted affordances, this chapter takes up how participants teach and learn cadaveric anatomy through observation, specifically, how they gain the ability to recognize anatomical structures in and on cadavers. Through the convergence of seeing and touching, participants learn and teach anatomy through hypothesis confirmation, namely a search for visual and tactile evidence that enacts the physical body as the anatomical body by allowing participants to recognize anatomical structures on display. By developing an awareness of how a structure should look (descriptive evidence) and its associations with neighboring structures and basic physiology (relational evidence), students recognize anatomy in the body by learning to recognize and to be persuaded by anatomical evidence. The cadaver, then, functions as a rhetorical object that not only presents anatomy but also appeals to the students' budding expertise. As I will illustrate, recognition and persuasion derive from the students' skillful ability to bring together vision and touch and to analyze perceptual content in light of anatomical knowledge. Observation, in this and other technical settings, offers a concrete example of embodied learning made possible by the merger of a haptic gaze and rhetorical processes of demonstration and verification.

Medical Observation as Embodied and Rhetorical

Scientific observation, according to Haraway (1988), is always "the view from a body" (589)—a particular embodied perspective from which a human looks out onto and even interacts with the world. The same is true for medical observation, though we persistently categorize it as a predominantly visual domain of standardized objectivity, thus downplaying the embodied perceptions of medical professionals. In her ethnographic study of surgical training, Prentice (2013) identifies a visual bias in medical education and practice that privileges both physical sight and the imagining of mental pictures. This bias, she contends, "erases the bodily, social, and relational ways of knowing" required for clinical education, diagnosis, and treatment (15). This overreliance on simplistic visual models, I would add, also ignores how all three ways of knowing are supported by and operate through rhetorical processes of demonstration, verification, and justification. In other words, the "medical gaze," as Foucault ([1973] 1994) terms it, is always an embodied and deeply rhetorical mode of perception.

In his investigation of eighteenth-century French medicine, Foucault ([1973] 1994) theorizes the nature of medical observation, exemplified by the gaze of the doctor or medical student onto and into the bodies of both patients and cadavers (108). This "perceptual act" (109) of observation unified the domains of clinical care and teaching in that the doctor's act of recognition, in and on the bodies of patients, paralleled the medical student's "effort to know" about the body (110). Yet the results of these observations were often ambiguous, forcing students and teachers alike to reason out and deliberate their perceptions and impressions. The use of spoken discourse, namely the interrogation of others (patients and doctors), transformed the silence of the look into a more contemporary understanding of the physical medical exam. Patients were examined visually and verbally, and students were made to account for what they thought they saw and heard. These verbal interactions between teachers and students recall the more interactive pedagogical style of Renaissance anatomists like Vesalius (Carlino [1994] 1999; Cunningham 1997). Demonstration, verification, and justification played a major role in medical observation.

To practice medicine by way of this embodied and rhetorical gaze, practitioners could not rule out other forms of sensory information. As Lawrence (1993) has shown, medical training in England involved training all the senses to detect clinical evidence and communicating that sensory-based evidence to others. For example, the doctor "translated" patient accounts into "symptoms with professional and lay meaning," and physicians accomplished this by simultaneously translating their "own sensations into perceptions" intelligible to other professionals (155). In the pedagogical settings of classrooms and clinics, physicians and professors taught future medical professionals how to garner evidence from sensory channels such as vision and touch and how to communicate that evidence in a legible and useful form. This education in embodied observation involved learning how to see and touch as well as how to think through evidence and communicate findings to others.

Gradually, as Borell (1993) argues, "machine-based technologies" came to augment and slowly displace the practitioner's senses (245). As Reiser (1993) reminds us, this displacement was gradual and involved the collaboration of the technology and the doctor's sense (263). Medical technologies, from x-rays to diagnostic tests, augmented physicians' physical senses, enhancing their clinical judgment and "emphasizing investigation rather than reflection" (Borell 1993, 259). Sight, in Cartwright's (1995) view, became the dominant sensory mode of medical investigation through imaging devices like radiography and microscopy that dispersed the medical gaze across an array of disembodying technologies. As a result, perception became "unhinged from the sensory body" and distributed across a number of "institutional techniques and instruments" (82). In the process, medical technology, according to van Dijck (2005), allowed clinicians to detach "diagnosis from an embodied perception of symptoms" (84). But as ethnographic research into medical imaging practices demonstrates, viewing and interpreting medical imaging technologies such as PET images and CT scans are rhetorical and embodied practices (Dumit 2004; Saunders 2008).

In contemporary medicine, as Dumit (2004), Prentice (2013), and Saunders (2008) bear witness, clinicians must contend not only with visual inscriptions of symptoms, illnesses, and bodies, but also they must recognize, interpret, and deliberate about these inscriptions while using them to communicate their evidence or conclusions to others. The same is true in the gross anatomy lab, where learning anatomical knowledge and skills involves more than simply seeing and knowing. To learn anatomy, participants must supplement visual evidence with other forms of sensory evidence, particularly clues derived from haptic, or touch-based, experience. Also, participants—instructors, TAs, and students alike—learn to couple visual and tactile investigation with verbal interrogation and reflection. However, unlike living patients, cadavers are poor interlocutors; therefore, anatomy students cannot verbally interrogate the bodies before them. Students can and do interrogate each other, the instructors, and the TAs because anatomical objects, specifically cadavers, are not always what they seem. From these observations and interrogations, students learn to reflect on anatomical evidence on display in a process that trains their perspective, actions, and judgment.

Anatomical Evidence: Descriptive and Relational

To assess anatomical evidence in a cadaver's body, participants must look at and feel the structures on display and confirm what they think they are seeing and touching. To do so, students transition from what one instructor terms "photographic anatomy" toward "evidence-based anatomy." The first stage of this process, photographic anatomy, is largely a visual endeavor:

> [The student's] brain takes a snapshot, right? And then when they're asked, the brain has a snapshot of things they think they can associate with a name. And when they take the exam, they go through their snapshots, and they match it, and give it a name. (William, instructor of both courses)

These "snapshots" of the body produce what the instructor terms "photographic anatomists" or knowledge of the descriptive evidence of the structure in question. (Again, "structure in question" is my way of designating the specific anatomical structure participants are searching for or working with at any particular moment.) By "descriptive evidence," I mean the physical and visual description of an anatomical structure in relation to its observable external or internal characteristics. Throughout the semester, the instructor warns students not to stop at the level of photographic anatomy. One afternoon in the prosection course, he provided a memorable anecdote about the limitations of "photographic" evidence.

"Last year we tagged a pancreas for the exam, but everyone thought it was a penis," he says, without cracking a smile. "They saw something tagged sticking out and thought it was a penis." Shrugging his shoulders and raising his hands in a gesture of mock confusion, he mimics a perplexed student: "If something sticks out, it must be a penis, right?" There is a bit of laugh. He continues by telling them not to be a "photographic anatomist." That is, they should not think "if it looks like the picture, it must be it." If they do, they are not really learning the structure; instead, they are learning how the structure should look. As this illustration makes plain, learning anatomy involves more than visual memorization of an object's appearance. The instructor's use of the word "snapshot" is noteworthy in that it emphases the visual textuality of the body as an image, a picture of anatomy. Students must take great pains to move beyond this image if they are to learn. One way students transcend photographic anatomy is by getting in touch with—physically and figuratively—the three dimensionality of cadaveric bodies. After all, the physical human body offers a complexity that 2-D anatomical snapshots lack.

Surpassing this knowledge of mere appearances involves what the instructor of the course terms "evidence-based anatomy," and what I call "relational evidence." His term alludes to the current movement known as evidence-based medicine (EBM), a push to "eschew unsystematic and 'intuitive' methods" of patient care in favor of "a more scientifically rigorous approach" (Goldenberg 2006, 2,621). EBM seeks to make clinical research and deliberate decision making the foundation of medicine, thus downplaying the subjective art of medical practice. By structuring patient care according to standardized guidelines drawn from clinical research findings and other verifiable evidence, proponents of EBM seek to enhance health care by increasing evidence-supported procedures across a number of clinical settings (Berg 1997, 1,082).

In the gross lab, the evidence on display does not originate in clinical trials or standardized guidelines. Instead, recognizing and appreciating anatomical evidence means understanding the relationships between structure and function as well as a structure's relationships to neighboring structures, many of which serve as landmarks. Transitioning from initial descriptive evidence to more advanced forms of relational evidence entails a type of rhetorical interrogation in which a student posits a hypothesis and uses the evidence on display to arrive at the identity of a structure. In the words of that same instructor, William,

> I think maybe in your identification, [photographic anatomy] is always your first step. Then, once you move from that step to the next, which is evidence based, you ask yourself, "O.K. This looks like it." And then you look at all the information around and say, "Does all the peripheral information convince me that it is this?" (William, instructor of both courses)

This second stage, the recognition of the relational evidence of structures in question, entails analysis of the visual and haptic features on display in cadavers. His use of "convince" is telling and typical of the participants I interviewed. The students, TAs, and instructors all used rhetorical language such as "persuade," "convince," and "argue" to describe the movement from the visual memorization of anatomy toward the confirmation of anatomical knowledge. Weighing relational evidence involves rhetorical verification and internal persuasion.

Participants come to recognize and weigh anatomical evidence through a type of haptic gaze, a physical and metaphorical sight augmented by the sense of touch and other embodied perceptions (Prentice 2013). This way of seeing and feeling, I argue, is quite different from usual formulations of clinical judgment that physicians and other health professionals use to diagnosis and treat the patient-body. Montgomery (2006) conceives of clinical judgment as akin to *phronesis*, or practical reasoning, rooted in a "narrative rationality" that is driven by the medical case narrative on which physicians rely to deduce clinical causation (46). That is, the goal of clinical judgment is the search for causes, using medical knowledge and practical reason to diagnosis patients. Clinical judgment, Montgomery suggests, begins with the doctor's question to the patient: "What brings you here today?" (61). From this search for causation, clinicians enact the patient-body.

Anatomical observation, on the other hand, is an ontologically directed process of hypothesis confirmation that uses anatomical evidence, both descriptive and relational, to find or recognize anatomical structures. The key questions here are ones the lab participants ask themselves: "Is this what I think it is?" and "Am

I seeing what I think I am?" Keeping in mind stasis theory of classical rhetoric, I contend that anatomical deliberation corresponds to what Prelli (1989) has termed "conjectural stasis," which "occurs in evidential scientific discourse whenever there is ambiguity about the available or reliable evidence" (148). To adapt Prelli's words, the general question becomes "Is there or is there not evidence to support the claim that this structure in question is what I think it is?" (149). In the gross lab, these questions set in motion a rhetorical process of demonstration, verification, and justification that articulates the anatomical body (of the lab) as a precursor to the patient-body (of the clinic). After all, physicians must learn to recognize the anatomical body before they can use this knowledge to deduce symptom causation from the patient-body. If they do not understand what they are seeing in the gross lab, they cannot understand what might be wrong with similar bodies in clinical settings. Anatomical observation and the larger practices of trained vision necessarily precede the development of clinical judgment. As I demonstrate in this chapter, anatomical observation also trains students in similar rhetorical skills that are required to make diagnosis.

Anatomical Observation and Bodily Skill

In the gross lab, participants learn physical and rhetorical skills necessary for embodied observation, a way of seeing, touching, and thinking that facilitates learning and contributes to the development of expertise. Through vision and touch, specifically the recognition and verification of descriptive and relational evidence, participants use their bodies and the objects of their environment as perceptual tools for experiencing and making sense of the lab. Dreyfus (2005) has sought to explicate the process of skill acquisition by exploring how our "relation to the world is transformed" as we acquire a skill (130). To conceptualize a working model of how the body facilitates learning, Dreyfus argues that we acquire skills "by dealing repeatedly with situations that then gradually come to show up as requiring more and more selective responses" (132). In other words, humans, in a specific social setting, repeatedly encounter tasks that progressively require more and more skill. More successful participants learn to master the primary tasks and move on to more complex ones. These successful learners eventually perform these tasks without a need for conscious thought. The tasks and the skills used become second nature to participants who acquire the skills needed to move on to more complex tasks. For example, experienced drivers do not have to consciously consider the tasks and bodily actions needed to start a car and back out of the driveway; they do it without a conscious acknowledgment of the steps they once learned so deliberately.

Dreyfus calls this process "skillful coping," and he bases it on Merleau-Ponty's ([1945] 2005) notion of "the intentional arc," defined by Dreyfus as a type of "feedback loop between the learner and the perceptual world" (Dreyfus 2005, 132). Merleau-Ponty ([1945] 2005) understands cognition and perception to be "subtended by an 'intentional arc' that projects round about us our past, our future, our human setting, our physical, ideological and moral situation" (157). Phenomenologically speaking, to say that something is intentional is to say that it "aims toward" something else; intentionality is, in Thompson's (2010) apt phrase, "object-directedness" (22). For Thompson, the intentional arc is "a body-environment circuit" of perception and action, comprising objects, displays, documents, and discourses, plus the sensorimotor activities necessary to make sense of those objects (248). According to this skillful coping model, the learner's past experiences are, in Dreyfus's words, "projected back into the perceptual world" and appear, in a Gibsonian sense, as "affordances or solicitations to further actions" (2005, 132; see Gibson 1986).

When anatomy students demonstrate, observe, and dissect, they are enmeshed in an intentional arc of activity and meaning making. They are repeatedly engaged in activities that require them to use their entire bodies in iterative ways. To succeed in learning anatomy, students must learn to cope with the perceptual tools of that arc—tools that include the lab's displays, documents, objects, discourses, and their own bodies. In the process, students not only learn anatomical knowledge but also internalize certain skills, tasks, movements, and perceptions as second nature. These second-nature activities contribute to and help constitute their expertise, their trained vision. As a person becomes an expert or develops a certain trained vision, "the world's solicitations to act" in specific contexts and under specific constraints replace conscious models and representations of what to do and what not to do (Dreyfus 2005, 132).

Learning, according to this phenomenological approach, is a complicated process in part because participants face a host of objects, choices, and tools that constitute the intentional arc. In these situations, the learner strives to achieve, in the words of Merleau-Ponty and Dreyfus, "a maximum grip" on all components of that activity—the displays, objects, documents, and discourses of the environment (Dreyfus 2005, 137). By learning to use our bodies in particular ways, we coconstruct and come to grips with the world around us. Again, as Noë (2006) and Thompson (2010) contend, perception and cognition are accomplished not just through brain-bound processes, but instead through the activities of the body engaged with the world around it. We know and enact the world through our bodily perceptions, which are forms of action (Noë 2006; Thompson 2010). Thus the body, through embodied learning and skillful coping, is the means by which we understand the intentional arc of activity. To learn a skill, we must attain a maximum grip on the intentional arc of activity, which includes the displays, objects, documents, and discourses around us. In seeking this grip on the resources and tasks that surround us, we, as skillful copers use our body as both an instrument and an experiential space. Skill acquisition requires more than learning and applying rules; it requires mastering the opportunities and objects of a domain so that we might apply the rules or use the objects in a seamless manner (Benner 2001; Dreyfus and Dreyfus 1986). At that point, the rules and objects become second nature to the hands, eyes, and bodies immersed in the experiences of skillful coping. As Merleau-Ponty ([1945] 2005) describes it, the body has learned this skill in such a way that a habit "has been cultivated" (169). That habit is not as much an unconscious tendency as it is a "form of responsiveness to the environment" whose mechanisms the human no longer has to consciously consider (Noë 2009, 127). My aim in this chapter, and throughout the book, is to make plain the role rhetorical practices and discourses play in these cognitive, phenomenological, and embodied "forms of responsiveness."

Learning anatomy through observation is an embodied process of skillful coping not simply because participants use one body to learn another, but also because they use their own bodies and their developing expertise to make sense of, or come to grips with, the subject matter, the tools, the objects, and their own perceptions. In the gross labs, this "grip" is not just metaphorical; it can be a literal, tactile hold on the physical human bodies on display. To be successful, students must make choices and use certain resources, including the haptic gaze needed to confirm anatomical values. Further, they must receive feedback, positive and negative, from that intentional arc, which may take the form of others' comments or an internal feeling of correctness. Students achieve this sense of internal confirmation by learning to distinguish between looking at cadavers (and visualizing structures) and knowing the anatomical evidence at their fingertips.

Looking versus Knowing: Grasping Evidence

Due largely to the abundance and variety of multimodal displays students encounter (and will use as future medical professionals), understanding the descriptive evidence of structures, visualizing their location and physical appearance, is the first step in learning anatomy. After all, the images and objects offer schematized, idealized, or exaggerated representations that influence how students visualize the anatomical body. As a result, students easily and understandably come to focus their attention too heavily on mental visualizations of anatomy because nearly everything seems oriented toward that end. Even studying the terms can lead students to an unhelpful dependence on physical description because a structure's name can be embedded with visual clues, such as glottis for tongue or cranium for head (as one student pointed out). By using *Netter* plates to learn anatomical terms, students work with descriptive evidence. Visualizing anatomy and understanding descriptive evidence are necessary to the course because, as I will discuss in the next chapter, a photographic sense can aid dissection. Studying with *Netter* and working with descriptive evidence can guide a team's dissections, helping them become more adept at creating a 3-D cadaveric display.

For students in the prosection course, visualization practices help them make the most of laboratory sessions. By studying with these 2-D displays, they learn to "recognize anatomy" (as one student put it) in the cadaver. Yet an overreliance on *Netter* and any 2-D display becomes a problem when students fail to grasp the importance of learning a structure's landmarks, its relationships with other structures, and crucial nonvisual properties. One undergraduate TA, Kate, explains what she finds problematic about the tendency to memorize the look of structures:

> Because if you memorize that a muscle is looking a certain way, that may change anatomically. The variations you may have from cadaver to cadaver may be different. So it is more important that you know the landmarks that surround it.

For Kate, developing this particular form of medical expertise while learning anatomy requires more than a knowledge of physical description; instead, anatomical expertise develops from an appreciation of natural anatomical variation and, perhaps more importantly, a nuanced understanding of neighboring landmarks. Memorizing is part of learning, but the trick is to memorize properly, in Kate's words, "based on the landmarks and not because it looks a certain way." A reliance on descriptive evidence can be further complicated when the structure in question has been removed. If a student can recognize a structure only by its appearance, she might in confusion identify a similar structure if the one she is looking for is absent. TAs in both courses stated time and again that learning anatomy meant learning systems and coming to grips with how structures work together—evidence-based anatomy and the relational clues it reveals.

 Understanding relational evidence involves incorporating visual clues with more advanced anatomical knowledge, namely recognizing structural relationships and haptic evidence such as texture, depth, scale, movement, and other kinesthetic qualities. An awareness of the complex relational values of structures not only makes one a good anatomist but also constitutes a more reliable anatomical knowledge. One dental student offers an explanation of the layers and components of learning anatomy:

> [Students who understand anatomy] look at the whole puzzle and then, kind of, can flow through it. That's a good way to put it. Kind of look at it like it's a puzzle. Some people take each piece and then try to find it, or some people group pieces together and then lock it into the puzzle.

The student went on to say:

> Like people who can, I think, say, "O.K. I know this muscle is the stylohyoid because it goes through the stylohyoid, and it's near the hyoid. So I know where that is." And I think others just memorize it as stylohyoid, as a blank name, and they know exactly where it is. But they don't understand the relationship, and why it's called what it is. (Larry, dental student)

Using a jigsaw puzzle metaphor, Larry describes learning as a slow, visually oriented, and visually directed process of bringing together various components that are eventually made to fit into a coherent image. Yet, like any puzzle, the connections are already predetermined and students must understand where and how to make things connect. To do this, students follow the evidence on display, which, like a jigsaw puzzle, is not always visual. Whether assembling puzzle pieces or learning anatomical relationships, participants use touch and vision to put the larger picture into place, judging with eyes and hands what pieces fit where. Because Larry knows the structures' names and recognizes their location in relation to landmarks, he can fit the pieces together. The relationships are crucial here; they give coherence to the mental visualization of structures.

Though students can use repetition and memorization to learn names and landmarks, participants explain learning as a process of either appreciating the three dimensionality of the anatomical body or working to three dimensionalize any human body. This discussion below, from a medical student, illustrates how relational evidence and three dimensionality (two components of Larry's puzzle) are not always transparent:

> Up until the first test, I was studying so hard I felt the material just wasn't sinking in—until I realized that I kind of needed to come at it from two different directions. So name the nerves and all the branches and what they do, and then go back and name the muscles and what nerves innervated them. (Ramona, medical student)

Ramona, through a self-assessment of her performance, comes to doubt her study habits because of questions she has about her comprehension of anatomical knowledge. Though she does not link her understanding of the anatomical body directly to touch, she does use embodied, figurative language to describe what it is not doing, namely "sinking in." Her use of figurative language of three dimensionality is significant, but more than that there is her awareness of the need to "come at it from two different directions," a spatial analogy describing movement. For her, those directions incorporate knowledge about two sets of related structures that work together in the body: the nerves and the muscles they innervate.

Ramona's ability to see or visualize the body in three dimensions helps her learn anatomy by identifying how structures work together. This awareness is complicated, however, by the course's multimodal displays:

> I am just trying to picture everything as a whole, how everything fits together. The biggest problem is just picturing everything in three dimensions. And having the body there is definitely a lot of help, even though it is usually supine, in a supine position, or in the pronate position. So you still almost visualize it in just two planes. (Ramona, medical student)

Here she directly expresses the importance of 3-D relationships, but then complicates that by pointing out the limits of cadaveric bodies. As lifeless flesh lacking motility, cadavers cannot easily be moved or manipulated outside of their resting, supine position. The three dimensionality is there, but participants have to visually (and physically) work to conceptualize it. Thus, even when participants acknowledge the significance of relational evidence, they must grapple with the material limitations of the cadaveric displays in order to perceive that evidence.

In what remains, I turn to a deeper explication of data to demonstrate that learning and knowing evidence-based anatomy, or relational evidence of the anatomical body, is only possible if participants learn to physically interact with the lab's various bodies. Specifically, instructors, TAs, and students must physically and rhetorically interrogate those bodies and their own in order to convince themselves that they, in fact, know what they think they see. In the process, participants continue the development of trained vision and technical expertise that embodied learning and rhetorical demonstration make possible. In the gross lab, observation and embodied learning are comprised of a haptic gaze and rhetorical demonstrations easily visible in five common laboratory activities: (1) performing dissections, (2) teaching prosections, (3) studying in groups, (4) participating in clinical-correlation sections, and (5) tagging cadavers for exams.

Doing Dissections: Digging through the Body

To dissect a body, students literally excavate layer by layer the various outer structures in order to arrive at inner ones. As I mentioned in the previous chapter, participants first learn this process of excavation on the 2-D, representational objects, usually naturalistic displays like *Netter*. In the process, students learn to relate one multimodal display with another:

> I take *Netter*, and I just dig through *Netter*. I go over and over and over. I have a photo atlas too that is actually pictures. I go through that too. And I just see what it all looks like. I try and get it all as memorized as possible. So that I can get to where I can just look at the structure and identify it. I do that before I do it on the bodies. (Lynn, medical student)

Lynn uses interesting language: she digs through the book of photographic images before digging through the book of the body, in this case cadaveric specimen. By excavating or at least repeatedly viewing these lifelike displays, she projects anatomical terms onto these bodily structures. Lynn also attempts to use the photos to memorize the appearance and location of each structure. Later, in the labs, she will use this memorization technique to help her recognize structures in the 3-D bodies on the table. Even so, Lynn admits that learning to visualize a photograph's pictorial image is of limited value because cadavers are not visually identical to these professional photographs of expert dissections. Still, students and TAs praise the value of repeated exposure to naturalistic displays as a type of tacit, immersive, and indirect learning—a process that aids the more hands-on aspects of direct, embodied observations of cadavers. For example, one TA in the dissection course, Allen, recounts the advice he gives to students: "Just dissect and [do] not worry about the learning until you come back in [the lab for independent study period]." The dissection sessions, he acknowledges, can be loud and not the best place for studying; therefore, he encourages students to just "dig away and focus on that."

Most participants understand actively viewing and physically touching cadavers to be a major pedagogical benefit of dissection. One medical student, Barry, was interested yet skeptical of the non-dissection-based method used by another medical school's gross anatomy labs. These lab courses, based on demonstration and observation only, exclusively displayed already prosected bodies.

> So you can get the really standardized thing, but you miss the experience of digging through lots of fat to find particular structures that you know you need to see. Yet I think, like I say, there is a lot of variability. I don't know. Well, I think arguments could be made both ways. (Barry, medical student)

For him, dissection, here described as "digging through" the body, constitutes the learning and causes him to rethink his previous criticism of his own anatomy program's dissection model. Barry's ideas mirror those of other students who explained how well they understood a certain structure or a particular day's dissection because they were the ones to find those structures themselves.

Here are the words of a dental student, Marianna, who explains the relationship between their dissections (as processes) and the already prosected bodies (as products and models):

> Doing something myself, you know, actually helps me to remember it much better. But as far as wanting to know what structure is where, I will go to the prosection, just to make sure what I am looking at on my body is what I actually think it is. So I use the prosection as a major reference. (Marianna, dental student)

Again, dissection is a process of active learning, of viewing and touching, but the prosection also offers an opportunity for learning, one that on the surface seems to be a strictly visual process of looking and figuring out. But, as Marianna continues, we see that this is not the case:

> I think that's what a lot of people do. Because I have noticed that when we have prosection, people always come over and look down at it, tinkering

inside it, and then go back to their tables. And then they say, "Well, we need to cut here," and that kind of stuff. So yeah, it helps them both in identifying the structures and in the dissection process.

Viewing prosections also involves haptic evidence. By "tinkering inside" the prosected models, students realize depth, orientation, and arrangement, features that guide the students' dissections. Though there is obvious anatomical variation, students use knowledge gained from one body to understand and learn from another. However, this analogy-based comparison is only possible if students understand the relational evidence of the prosected body and the body they are dissecting. Because prosected models provide a typical example of how the students' own dissections should eventually look, making sense of prosections requires photographic and evidence-based considerations.

Teaching Prosections: Demonstrating Bodies and Knowledge

More than just cadaveric models, these prosections serve a central role in the development of the haptic gaze and the rhetorical processes of demonstration and interrogation. More specifically, the brief prosection demonstrations invite students to become not just spectators to an exhibition of evidence, but more importantly, active interrogators of meaning. Teaching with and from a prosected cadaver is challenging because students must assume teaching roles with multimodal objects they may not fully understand. Take, for example, this common scenario provided by the medical student, Mitch:

> I get the most out of [prosection demonstrations] when I do the prosection demonstrations to people. We show up. It is a very cleanly dissected. All of the fat and the garbage is out of the way. And [the presenters] teach us step-by-step what comes out of this opening, what goes where, and that kind of stuff, which is helpful. Otherwise, it is just four idiots who don't know any more than the other one. It's like the blind leading the blind.

Though his words might seem a bit harsh, Mitch voices a widespread complaint: dissection can be a frustrating even distressing process because of the unruly cadaveric body. This frustration is exacerbated when students are instructed to guide others through the body and through the process of dissection. His mention of "the blind leading the blind" indicates what can happen if students performing the demonstrations are unprepared. Without previewing that day's material, those four students may receive their first and only introduction to those anatomical structures during the 15- or 20-minute practice presentation with their TA. These unprepared students then have to demonstrate those structures to their classmates. On another level, his reference to blindness implies the use of touch and tactile sensation to compensate for a lack of visual clarity.

Another complication of prosection-based teaching is the language students use in demonstrating to each other:

> Some groups [who act as prosection demonstrators] do an awesome job, and other groups just kind of fly through it and just want the next group to come over. So it really helps when some groups go slow and show you the arteries and everything, where they branch off of, and not just point at the artery saying, "This is an artery," or something like that. (Amelia, dental student)

Amelia alludes to a "this-is-that" type discourse to which unsure or uninterested demonstrators resort. Throughout the dissection course, students and TAs disparage these reductive verbal descriptions because of their dependence on context-specific pronouns and photographic anatomy. Knowing only that "this is that" in one particular body does not help students understand how to find "this" in another body because they become reliant on finding a structure that resembles "that."

Instead, TAs and instructors encouraged students, as one instructor put it, to "teach the body, not point at it." Dental student Marianna knows exactly how she wants to be taught:

> I want a relationship, and I want compartments. Like I want, "Here is this artery," and you can find three branches in the anterior compartment and three branches in the posterior compartment. I don't want them to just go through and point and name what each thing is because that doesn't help [me] learn.

What she wants to know are relational values, "compartments" and "relationships" as well as the ways that structures work together. Comprehending and recognizing branches of arteries offer a great deal of information about location and relationships. In effect, the descriptions students deem most helpful are ones that narrativize the body: "We say the structure, and we go from there, do a little story about it. And for us, I think that really helps" (Amelia). These narratives involve ascribing a kind of bodily or kinesthetic agency to the structure in question. Nerves, veins, and arteries "run" and "dive." Muscles at times "pull" instead of contract. Some structures "hide" behind others. These narrative accounts provide relational evidence by focusing primarily on the physical location of structures, that is, where they run, more than their physiology, or how they function. These anthropomorphic descriptions, in a sense, translate the position, location, and landmarks of structures into actions that are easier to visualize (recall) and recognize in another cadaveric body—not to mention far more precise than repeatedly pointing and saying "this is that."

Learning to illustrate and narrate the body's relational properties is important because students perform prosection talks throughout the semester, during official labs and unofficial study periods. Students who understand the relational complexity of the anatomical body and who can communicate that knowledge in a way that preserves the interconnectedness of structures and system possess the skills necessary to teach themselves and their peers. This ability to view the body in a certain way and then communicate that understanding allows students to assess their comprehension of anatomy, by comparing what they know to the prosection presentations, and their skill at dissecting, by comparing what they are told about the prosections with their own cadaver.

Studying in Groups: Interrogating the Body Together

Similarly, the third set of practices that help facilitate this haptic gaze are the group study sessions that students independently engage in during open labs sessions. These working groups, some of which are more formal arrangements created by students who want a level of consistency, offer a more overtly self-persuasive and interrogational learning. In both classes, though perhaps more so in the dissection course, students often study together, by either working with another student who happens to be in the room or with friends enrolled in the course. These groups of two to five students use their identification (or ID) lists and laboratory notes to perform as many demonstrations as possible. By moving from cadaver to cadaver and using other multimodal displays and objects, students teach each other what they know about the anatomy of each display, quizzing each other as they go:

> So one person will point at something, and one person will identify what that structure is. So it is a similar type of feel to the test because you have to come up with a name off the top of your head, rather than looking at the word on a list. (Randy, dental student)

As Randy states, these study sessions involve learning anatomy by mimicking knowledge necessary for the exams, which test students' knowledge by providing tagged structures they must correctly recognize and name. Learning names and structural relationships, usually through narrative accounts of what structures do (how they behave in the body), becomes a part of the interrogation process, encouraging students to share with each other all they know about the structure in question.

Students and TAs praise these intense, yet relaxed sessions because they allow students a chance to learn from cadavers and each other: "So if you are someone who needs that dialogue, and to talk about things, and to bounce ideas off other people, I think that is really effective for some people" (Stacy, medical student). As Stacy mentions, students work together to come up with answers (confirmations of anatomical structures) they might not understand on their own. During a typical session, a group of students will walk up to a cadaver and, either using their notes or working completely from memory, one student will play a TA-like role and begin asking others to identify and discuss a specific structure. Students take turns responding to this surrogate TA's questions. If there are points of confusion, disagreement, or complete misunderstanding, students will work together, offering clues, explanations, and suggestions. These moments of confusion allow everyone to assist, not just the student playing the TA. The team only moves on to the next set of structures when everyone in the group has understood. The interaction of these sessions involve not just the question and answer format of the this-is-that discourse, but more importantly the narrativizing of the anatomical body, as students require each other to back up their conclusions with some kind of evidence. In working out the identity of structures, students are required to offer up evidence based on anatomical relationships, often deemed more convincing than descriptive evidence that all present can see (and perhaps disagree with). During a study session in the prosection course, three students debated what turned out to be an artery and not a vein. In order to prove her point, a confident student asked her more skeptical companions to "reach in and touch it." She confirmed her hypothesis and persuaded her teammates by encouraging them to feel the structure in question and the structures around it.

These verification and justification activities mimic the way TAs and instructors quiz students during regular laboratory sessions. These exchanges, referred to by some as "pimping" the students, occur when TAs and instructors question students about a cadaver's descriptive and relational evidence. The quizzes usually begin with what the teacher or TA might consider more basic aspects of anatomy. But soon the questions get increasingly more difficult as students continue to answer them correctly. Instructors and TAs use these often rapid-fire question sessions in order to, as one undergraduate student put it, "kind of get your brain working a little bit more and kind of help you out as far as determining relationships." Prentice (2007) has observed that such quizzing activities encourage students to study "anatomical and procedural knowledge continually" as well as "keep on their toes" (547). These friendly though intense interrogation sessions happen in small groups; quizzed students thus exhibit their knowledge to anyone present. As public performances, pimping sessions mirror the quizzing rituals that Saunders (2008) observes as commonplace in radiological residencies. This "hot seat role," as Saunders describes it, inevitably "challenges and often embarrasses" the students while instructing them in the "rhetorics of testimony" indispensable to medical case presentations (205).

Though the use of the term "pimping" in this context might stem from the idea of "pumping" students for answers, the word intimates how participants instrumentally use one body for the benefit of another, while also reinforcing, as Prentice (2007) points out, the status and power differential between student and teacher (547). In its original use, a pimp (usually a man) forces a prostitute (usually a woman) into a brutal sexual economy in order to use the prostitute for the pimp's economic gain (and cultural capital). One body (the prostitute's) benefits and serves the purposes of another body (the pimp's) by serving the needs of other bodies (the customers', or johns'). In the gross lab, TA and instructors pimp students by forcing them to use cadaveric bodies to explain in stand-and-deliver fashion what they know about the anatomical body.

Metaphorically speaking, the docile body of the cadaver is not the stand-in for the prostitute; instead, the quizzed student occupies that role in this perhaps troubling analogy. That is, students must demonstrate, even perform, their anatomical knowledge on the body of the cadaver for the benefit of anyone standing around. In this way, the quizzed student uses the cadaver to benefit the onlooking students to the satisfaction of the instructor or TA. By verifying the physical presence of anatomical structures in cadaveric flesh, the student exhibits or performs the embodied knowledge and rhetorical skills necessary to recognize and demonstrate anatomy. This knowledge and these skills, in turn, inevitably bear witness to the student's more advanced expertise. Cross-examination, rather than pimping, might be a more fitting term because it describes the interrogation of a person in order to provide evidence for a larger group of witnesses.

Applying Clinical Correlations: Incorporating the Body

The bodily analogy implied by the term pimping as a description for the interrogation of students intimates other ways students use their bodies for the benefit of others. Here I am referring specifically to the way students come to incorporate anatomical knowledge by either projecting it onto living bodies or performing certain motions to make that knowledge more visible. This awareness of clinical correlation is the fourth set of practices dependent on embodied observation. This use of gesture and physical demonstrations illustrates a more advanced, evidence-based anatomical knowledge:

> I think, obviously, [we] use the cadavers, and we use a lot of our own, you know, our own body to point at things because sometimes it is hard to, kind of, sometimes conceptually see things because you can move an arm while the cadaver can't move its arm for you or for itself. And so we use a lot of our own, I guess, body parts to point to. (Constance, dental student)

According to Constance, cadavers and living humans offer a demonstration of some structures or processes, thus displaying the anatomical body. But the limitations of cadaveric bodies can force students to work with living ones. In these cases, students move, turn, and manipulate each other's bodies in order to explain more advanced concepts like motion, origin, and insertion. Again, as Constance explains, "If you know the muscles, if you know its action, then you know its origin and insertion," and so "using the [living] body is best for muscles." These clinical insights can be difficult to grasp from inert flesh.

These clinical correlations, which reconnect anatomical knowledge to the lived experiences of the body, link structures to functions in a way that introduces students to "the whole puzzle" of anatomy. All the participants I interviewed found these correlations not only illustrative and interesting, but also motivating—stimulating these future medical professionals to really learn anatomy and not just memorize it for exams. These correlations involve the same processes of interrogation and self-persuasion witnessed in quizzing and group study. Take Stacy's account of her own personal connection to these correlations:

> It also just makes it more interesting. It is kind of like another one of those "O.K. people, if we cut this then you're going to have this." And then you're like, "Oh, right." Like if you sprain your ankle—like I have played soccer, and I had played for 18 years. And I have sprained my ankle a hundred times, and it wasn't until we had our clinical correlate when [the TA] was like, "Yeah, when you have a first-degree sprain, your anterior talofibular ligament is like this." And then I'm like, "Oh." And that was on the practical [exam], and I was like, "O.K. That's the one I keep spraining." (Stacy, medical student)

Here we see one student learn anatomy by personalizing the process, physically incorporating the anatomical body by understanding her body in anatomical terms, thus expressing what I have called the reality effect of anatomical demonstration. However, more than enacting the anatomical body as the physical body, she recognizes the patient-body. What was once a sprained ankle is now an ailment caused by a knowable mechanism; this knowledge will assist her future clinical practice not to mention shape her relationship to her own body. Gross anatomy students (most preparing for medical careers) embraced anatomical knowledge, especially a keener awareness of structure–function relationships, as empowering and productive.

The physical application of clinical correlations, for the medical students at least, begins in the labs. During the dissection course, instructors organize and require medical students to participate in official clinical-correlation sessions geared toward connecting anatomical evidence to clinical judgment. During these sessions, medical students meet in small groups in tiny examination rooms; there they work with TAs or instructors to make explicit the relational evidence of anatomy for clinical practice. Usually by palpating each other, students learn how certain structures work and how these systems comprise the human organism. For example, this passage comes from a medical student who found these sessions illuminating and entertaining:

> I had not totally understood the nerves yet, and in the clinical correlates they taught us different [physical] exams, you know, where to check, sensory innervations on the hand, and stuff like that. If you couldn't feel this

part of your hand, then it meant there was a problem with a certain nerve or something. So actually, it helped me put together where things were running because of the clinical correlations. We actually went into one of the little exam rooms and, you know, did stuff on each other. And, you know, that was really cool. (Jennifer, medical student)

By doing "stuff on each other," Jennifer was able to better understand a conceptually challenging system of structures—how nerves innervate muscles and how that innervation makes motor function possible. By admitting that she was able to "put together where things were running," Jennifer not only gives a narrative account of the body but also expresses the importance of understanding how structures work together, an understanding she gained by touching the body of others and moving her own.

Though during my year of observation dental students did not participate in these special sessions, the regular dissection course introduces all students (both medical and dental) to clinical applications of anatomical knowledge. Through these moments of exposure to clinical knowledge, more skilled learners are able to translate description and relational evidence into anatomical knowledge rich with clinical implications. One example is the donor medical histories printed out on each cadaver tank. This one-page list of that donor's pathologies and ailments offers students (and TAs) clues to what they might find (or not find) in the body. If a donor had her gall bladder removed, this donor history will explain the absence of that structure. When I asked one student whether he and his team ever looked at those histories, this was his response:

As far as the initial learning of the anatomy, you would be kind of like, "What is this?" [mimics noticing and reading the sheet]. But then you would figure it out, and then it would become really, really interesting, particularly after you had learned the basic anatomy. And then you see variations based on procedures that have been done or things that have gone wrong. It is just like, "Oh, wow, now I understand everything even more. What they did when they did a coronary bypass, or a gallbladder was removed, or this person was a smoker, what that looks like." (Randy, dental student)

He exemplifies what participants in the dissection course expressed: the donor histories are interesting and useful, but only "after you had learned the basic anatomy." And "then you see" relationships in the body and clinical implications and applications—the advanced anatomical knowledge of relational evidence. These relational values are a form of embodied knowledge that participants acquire through the haptic gaze and rhetorical demonstration techniques. The use of these donor histories and other forms of clinical correlations are inevitably a form of embodied evidence that is persuasive to students because it motivates them to learn anatomy and, as some have mentioned, take better care of their own bodies.

These clinical correlations bear witness to the patient-body's emergence. Though recognizing relational evidence means comprehending connections among a structure's look, physiology, and position, this evidence-based knowledge is focused nonetheless on learning anatomy itself. It is not directed toward diagnosis or treatment. Clinical correlations require participants to recognize these relationships in living bodies in order to understand how anatomical structures make movement, respiration, and other bodily processes possible. Yet, they also make visible the consequences of anatomical failures or the results of surgical procedures that remove or reshape the body's anatomy. Clinical correlations, then, reconnect the structure and function of the body to possible complications that might cause a person to become a patient. Students and TAs begin to enact the patient-body from the possibilities of the anatomical one, constructing the former from the material of the latter through a haptic gaze that brings together sight and touch, along with rhetorical processes of demonstration, verification, and interrogation. But again, students must understand anatomical and physiological knowledge before they can use that knowledge for diagnosis or treatment. In recognizing clinical correlations, students learn to identify anatomical relationships responsible for normal physiological functioning. Students develop this understanding by coupling rhetorical demonstrations with the haptic gaze.

Tagging for Exams: Weighing the Body's Evidence

The TAs of each course find themselves in a special pedagogical position that requires them not only to master these techniques of embodied learning, techniques that involve the skillful grasp of anatomical evidence, but also to teach these observational and rhetorical skills to students. As William, the instructor of the prosection course, explains, "One of the goals of being a TA is to become a higher level of anatomist." As such, he urges them to leave behind the undergraduate book and take up *Netter* because the ability to perform advanced dissections and demonstrations guided by these naturalistic displays will develop TAs' expertise, specifically, a sophisticated ability to recognize and teach descriptive and relational evidence. The TAs' unique deliberation on this evidence, made possible by embodied learning, is most apparent in their tagging of cadavers for exams. Tagging a cadaver involves the use of colored pushpins or labeled strings to mark structures so that students can identify them (without touching) during the exam sessions. A tag becomes a teaching tag when instructors and TAs use it to indicate a beautiful dissection or structure—a clean, clear, netterized, almost image-like view of a structure that seems to demonstrate not only the structure in question but also the evidence important to identifying that structure. By providing an emblematic example of that structure, a teaching tag is never ambiguous to participants who know anatomy; thus it reinforces anatomical knowledge in participants who can weigh the evidence properly and know what they are seeing. Interestingly, though course instructors differentiated "teaching tags" (ideal for teaching structures) from "exam tags" (ideal for testing structures), most TAs used the term "teaching tags" for both.

Tagging for exams challenges TAs to clearly tag expertly dissected structures in order to create challenging (but not impossible) questions. Identifying anatomy should not entail a process of trial-and-error reasoning: "There is no such thing as 'if it is not this, then it must be this' in anatomy," William explains during one of our interviews. Tagging structures requires a great deal of deliberation and confirmation; as such, TAs and instructors work together to decide what structure to tag, where exactly to tag it, how to orient the cadaver, and how to orient the tag. They also carefully netterize the area to create a well-exposed dissection, revealing enough neighboring landmarks and points of origin and insertion to make it "knowable" but not easy. Students should be able to deliberate using descriptive and relational evidence, not just descriptive anatomy. Instructors and TAs do this work by keeping in mind how a student might view the cadaver. For example, while setting up the first exam for the prosection course, two TAs deliberated the difficulty of a particular tag because one of them felt the answer would be perhaps too obvious for any student who leaned close to the body. While they could not guess a student's physical position while observing tags, they wanted to ensure that the difficulty level would not depend on how students viewed the cadaver.

When tagging structures, TAs and instructors consider how students might use all forms of evidence. This means TAs contemplate the evidence on display at a particular cadaver along with how and where to place tags to create a teaching tag. This requires that TAs, first, know and identify a structure in question by its descriptive and relational evidence. To identify it, they must know for themselves the structure's location, its major landmarks, and any structural and perhaps physiological associations crucial to identification. Second, TAs must be able to identify the best place to tag a structure in order to create a teaching tag. During one session of the undergraduate prep lab, two TAs, Mary and James, struggled to decide the location of a tag in part because they were unsure of the identity of the structure. Mary held a brain slice and told James where on the slice she wanted to place a pushpin to tag the structure in question. James told her how and where he would tag it, a response that differed from her suggestion. James explained that if he saw the tag in "that" place (her position), he would think it was something else. After he identified where he would place the tag, Mary raised a possible objection. This led them to *Netter* to double-check that they knew the structure's location, thus confirming their identification.

When the instructor entered, they asked him where they should put the pins. James pointed to a spot on the brain and asked, "Is that still [the structure] or is that not part of it?" The instructor took the brain and explained to them where the structure begins and ends, by marking both locations with a pin. At the end, he suggested they take forceps and "clean that section off a bit" to make the structure more apparent. Here, two TAs debated a particular tag by pointing out where the pins should be and demonstrating to each other the location of the structure in question. Because descriptive and relational evidence is more complex on an organ that can appear as one solid, undifferentiated object, the brain can be a difficult organ to tag, so TAs confirmed their knowledge through *Netter* and the instructor, who used this moment as an opportunity to teach them a more advanced anatomical point.

These rhetorical deliberations also occurred during the creation of exam questions, which are written on cards at exam stations. During a prep lab for the dissection course, a group of TAs and an instructor stood by a light box, debating where to tag an x-ray (using a colored arrow). The question centered on whether or not students would realize they were being asked to identify the second part of the duodenum, the descending portion of the structure that begins at the superior duodenal flexure. If they simply ask students to "identify the structure," would students (incorrectly) write "duodenum"? Or would students provide the correct answer: the descending duodenum? This quandary led to questions about the structure of the duodenum itself: was the descending duodenum really a separate anatomical structure or merely a section of the larger duodenum? Should students be expected to differentiate the two? Was there an important and, as one TA asked, clinical reason for knowing them as two separate structures? Nearly half the group felt that identifying the duodenum would be sufficient; those who disagreed were unsure of how to word the question. Would "identify" lead to two acceptable answers or would it confuse students? Would the phrase "identify, be specific" (one TA's suggestion) index the descending duodenum or would it just, in another TA's words, "trick students who weren't sure"?

Eventually, guided largely by the instructor, the team decided that medical and dental students in a dissection course should be able to identify specifically and correctly the descending duodenum, that knowing this structure not only evidenced anatomical knowledge but advanced observational skills. Though not specifically addressed by the group, the ability to identify this structure also bore witness to their ability to grasp complex anatomical evidence. In the words of the instructor, "You point out what is most clear and what is most correct, and not what would be nicest to them because that could then end up making it unclear to everyone." Ultimately, the group decided that the question should simply use the verb "identify," and the correct answer would be either "the second part/ section of the duodenum" or "the descending duodenum."

The objects and bodies that TAs, as advanced anatomists, must skillfully grasp include cadavers, multimodal displays, tools, and instruments. But TAs must also demonstrate knowledge of the students themselves, who become part of the TAs' intentional arc of activity. TAs must acquire the ability to assess the development of a student's embodied learning, anatomical knowledge, and technical expertise. TAs in the dissection course had perhaps a more difficult time of gauging students' ability to identify and differentiate anatomical structures through observation because the advanced course requires students to learn more complex, minute, and numerous structures. Though dissection is a part of creating teaching tags, tagging often means paying closer attention to how participants netterize a structure in question, as can be witnessed in the following scene.

One day in a prep lab for the dissection course, an instructor asked this question aloud to a small group of TAs in the room: "If you take the arteries out, have you taken out too many landmarks?" Looking down at her dissection of the cadaver, she surveyed the open body and considered her next cut. Two TAs, Betsy and Frank, leaned in for a closer look. Betsy described this identification as tricky, concluding that some students would not be able "to get that one without touching it" (which is not allowed in the exams). She thought it a little too tough to follow the landmarked structures around the actual tagged muscles. Another TA, Ruth, walked up and offered her opinion. "They can do that," she said, meaning that they could identify the structure in question. But she suggested leaving the arteries and removing the "confusing" neighboring structures. Frank felt this dissection would make identification difficult, but he added, "If they are careful, and if they have studied, they should be able to get it."

Consultation and confirmation are major components of dissecting the cadavers and tagging structures. During these consultations, TAs and instructors voice their opinions of what, where, and how to tag based on two factors: (1) their assessment of the evidence, and (2) their perception of what it is like to take the exam. In the earlier example, TAs disagreed over the difficulty of a tag based on their perception of the students' experience of looking or their own experience of looking. TAs and instructors pondered which neighboring structures provided the best view of relational evidence (and thus should be kept) and which were unnecessary for identifying a structure (and thus should be removed). These quick consultations, which TAs described as mimicking a medical consultation, allow TAs a chance to exhibit their knowledge and skill while assisting other TAs and the instructors, who often seek confirmation from TAs. Though instructors have the ultimate say, most decisions are made through consensus or majority opinion.

Because multiple practices are involved in tagging structures—deciding what, where, and how to tag, on which cadaver, not to mention what and how much to remove during netterizing—these consultation sessions are needed to double-check the clarity, position, and difficulty of tags. These tagged cadavers are, after all, the objects that instructors use to formally assess students' comprehension of anatomy and, thus, students' performance in the course. Obviously, there is an implied assessment of TAs as well; a tagged cadaver exhibits a TA's efforts in netterizing and tagging, offering supposed evidence of that TA's own anatomical knowledge, dissection technique, and deeper expertise.

During a prep lab for the dissection course, I watched a group of six TAs and three instructors move around the rooms, inspecting each tagged cadaver to double-check their success. They did this by successfully identifying the structure in question, thus confirming that each TA knew what the question was asking. The group began to debate the clarity and accuracy of a tag on one cadaver. One TA told the group, "I would know what that is"; two TAs agreed with him. One instructor paused before providing his assessment. He pointed to an area near the tagged structure, instructing them to "look at that and ask yourself what you would say that is." He subtlety implied that the tag would confuse students because it is unclear which structure is tagged. The TA who created the tag, Monica, looked visibly disappointed. The group then strategized what could be changed to make it more appropriate. Should they clean up the dissection a bit more or retag it in a different way? Eventually the group decided that both fixes were needed, and George, another TA, volunteered for the task, much to the relief of Monica, who had tagged the structure. After the group moved to another table, Monica thanked George, who reassured her that it was, in fact, a difficult structure to tag.

Though we see here one TA's failure to properly tag a difficult structure, I want to emphasize what it tells us about the difficulty of tagging and the complexity of anatomical observation, namely deploying embodied learning—specifically, the haptic gaze and rhetorical interrogations necessary to find, identify, and confirm descriptive and relational evidence. First, what made this tag apparent as unsuccessful were the instructor's questions concerning not the structure in question but the neighboring ones. When the TAs were asked to name the nearby structures, they began to see the tag as potentially confusing. Some students might have looked at the tagged structure and identified it immediately, perhaps through an overreliance on descriptive anatomy (the look of the structure). But for students who relied more on relational evidence to confirm their observations, this tag would be unclear because it called into question those relationships (their ability to identify neighboring structures). To identify anatomical structures and thus the anatomical body by way of observation, participants must not only engage in an interplay of looking, touching, and moving, but they must also use these embodied actions to deduce and debate what they see before them. For exams, structures must be tagged or rendered so as to maximize the students' ability to use embodied knowledge to recognize anatomical evidence. Often this involves the TAs in careful discussions of what, where, and how to tag as well as in last-minute dissections to netterize structures to make them more viewable (and, as I discuss in the next chapter, more beautiful).

This TA's attempt to create a teaching tag that makes viewable the structure in question, though perhaps embarrassing, is an inevitable part of all TAs' work. Tagging structures requires skills of embodied observation and the ability to mark those observations in a documentable form that renders them visible to others. Tagging also requires a kind of intersubjective awareness on the part of the TAs and instructors, as they must learn to view cadavers as the students might view them. To gain this view, participants must learn to see as another might see, to view from the position of another, and to recognize what might be persuasive to another. By developing this awareness, this rhetorical sensitivity, the course instructors and TAs experience and exhibit an intertwining of perspectives that allow them to render the anatomical body visible. They experience this intertwining by engaging in a more advanced from of observation, one that requires active reflection and self-interrogation. Tagging, then, is a rendering practice dependent on the embodied learning that participants develop and strengthen through their repeated exposure to the objects and practices of the lab—activities that constitute their intentional arc. Through habituated practices of observation, specifically the haptic gaze and rhetorical demonstrations, participants gain a more sophisticated grasp of these objects (cadavers, multimodal displays, tools, and instruments). Through this figurative and literal grasp, participants learn to identify anatomical structures on and in bodies, thereby learning the anatomical body of medicine.

Conclusion

Daston and Galison (2007) maintain that practices of scientific observation are "genuine technologies of the self" that mold observers' perceptions of themselves and the objects under scrutiny (234). Anatomical observation provides such a technology of the self in that it aims "to induce in students the habit of looking at the living body with anatomical eyes, and with eyes, too, at their finger ends" (Holden 1887, 1,025). To view the body in this way, hands must come to see as much as eyes, participating in forms of embodied learning that structure the deeply rhetorical practices of anatomical demonstration and verification. To learn, teach, and communicate anatomical knowledge, participants bring together visual experience (the act of looking), haptic experience (the act of touching), and anatomical knowledge (the language of anatomy) in order to identify the anatomy on display. These practices teach participants to juxtapose and even cross-examine multimodal objects with the knowledge and embodied skills they are slowly acquiring. By interrogating and physically interacting with cadaveric and living bodies, participants come to recognize and appreciate the descriptive and relational evidence of anatomy, that is, how structures should look and how structures relate to and work with others. Through this ever-growing awareness of anatomy's descriptive and relational evidence, students learn to incorporate anatomical knowledge as an embodied knowledge. To observe anatomy in and on the human body, participants form an intercorporeal relationship between observer and observed, in the process developing finely tuned physical and rhetorical skills.

This medical setting is not the only technical space where meaning and communication merge with vision and touch. In nearly every domain of practice, the development of expertise depends on a haptic gaze and rhetorical interrogations of people and objects. From computer programs they must master, to the experiments they must design and perform to understand how a machine will work or how must stress a material can withstand, engineers also learn by interacting with the technical objects of their discipline. Science students engage in laboratory experiments in part to gain an embodied and procedural knowledge of the inner workings of concepts they read about in books. Yet learning programs and conducting experiments are not transparent processes a practitioner performs only once with complete comprehension. Acquiring expertise as a scientist or engineer requires the internalization of not just knowledge and concepts but also the bodily skills and physical ways of seeing necessary to make sense of and replicate that knowledge. Technical and communicative mastery also requires the ability to demonstrate success, to verify results, to interrogate error, and to justify reasoning—all of which, as Graves (2005) has shown, require rhetorical deliberation.

The same is true for technical communicators who rely on active, physical interactions with the objects and processes they document. Hovde's (2001) study of how TPC professionals gain subject-matter expertise found that two of the most reliable ways to gain advanced knowledge of software was to use it repeatedly and talk to other professionals who do (86). Medical writers, in particular, rely on firsthand observations of the devices they document or the surgical procedures they describe. For instance, the anatomy labs I observed hosted special anatomical demonstrations of surgical procedures on human knee joints for the benefit of a medical device company's technical communicators and sales representatives. During sessions paid for by the medical device company, TPC professionals watch anatomy professors engage bodily with anatomical bodies. In the process, these communicators view and sometimes experience for themselves the haptic gaze and the rhetorical processes of demonstration and verification necessary to make sense of these anatomical objects and procedures. Standing around a dissected cadaver as an instructor provides a demonstration, the TPC professionals listen, lean in, touch, ask questions, and offer hypothetical scenarios to determine their own level of comprehension. Learning medical objects and bodies, both their descriptions and their complex relationships, is particularly importantly to these communicators whose work it is to produce the displays and documents on which others will base their expertise. By simply engaging in the work of their profession—traversing that intentional arc of activity— scientists, engineers, and technical communicators all engage in embodied rhetorical actions that connect them with the objects of their world.

5

MAKING BEAUTIFUL BODIES

Dissection as an Ordering Practice

The practices of any technical domain, medical or otherwise, involve the creation of orderly "science" out of what we take to be disorderly "nature." As Latour and Woolgar ([1979] 1986) explain, this desire for order and the elimination of disorder seems inevitable in any domain of human activity, whether science, politics, or ethics (251). But as Foucault ([1970] 1994) reminds us, the process of ordering is not a neutral endeavor; instead, ordering involves creating systems that we take to be natural, logical, or unexceptional. These ordering processes frequently go unanalyzed but exert a socializing force that inevitably shapes the perceptions and actions of those involved (Longo and Fountain 2013). This process of shaping perceptions and actions by way of ordering practices plays a significant role in laboratory work, where the "material operation[s] of creating order" include performing experiments, generating data, and writing up results (Latour and Woolgar [1979] 1986, 245).

As Lynch (1985) describes it, practices of ordering require the transformation of "specimen material" (37)—or the objects under investigation—into observable and analyzable data by way of "rendering practices" that make science and the work of science expressible and legible (38). These rendering practices are systematizing efforts such as marking, indexing, graphing, and mathematizing that normalize and even "civilize" the object under scrutiny (44). In the process, participants convert this specimen, which in medicine or biology usually means bodily matter, into a "docile object" (Lynch 1985, 43). These supposedly docile objects of science and medicine, however, often resist these normalizing efforts. Thus, the ordering processes of science are best understood not as creating order from disorder as much as reducing, though never eliminating, the "recalcitrance" of the object under scrutiny (Lynch 1985, 44).

Often these rendered objects take the form of visual inscriptions such as charts, graphs, or other displays of data and evidence (Lynch 1985, 51), and these inscriptions inevitably train the vision of scientific novices to view objects in a certain light. In the gross lab, rendering practices—cutting into the body (dissection) and presenting structures or images of structures (demonstration)—train participants to view bodies as instantiations of anatomy, and they do so by training the body to respond in a particular way. The experience of touch, the development of manual dexterity, and the embodied experience of simultaneously working on and being a human body: these contribute to the gross lab and the rendering practices that mark, systematize, and index physical bodies and multimodal objects. This rendering occurs through the interplay of cadavers and other multimodal displays, each used to make sense of and order the other.

During dissection, students and TAs manually maneuver through the physical human body, recognizing and revealing structures and transforming 3-D physical bodies into multimodal anatomical displays. Through a dual process of revealing to learn and learning to reveal, participants make the anatomical body visible all the while causing the physical one to slowly disappear. To dissect a cadaver in order to reveal structures is to make the body into a multimodal display by sculpting it, in a sense, to resemble as much as possible an illustration from an anatomical atlas. In the process, participants adopt an aesthetic orientation—an anatomical-aesthetic sensibility—that allows them to view the well-dissected cadaver as a "beautiful body," made so by its ability to become an authoritative presentation of real anatomy. In this chapter, I illustrate this process by way of field notes, interviews, and a visual analysis of von Hagens's *Body Worlds* exhibit of plastinated human cadavers, an exhibit that—for most participants in my study—epitomized this seemingly contradictory aesthetic category. Expert dissection renders the unruly cadaver beautiful and clean, and it does so by encouraging dissectors to use idealized atlas images as an interpretive framework—what Hutchins (1995) calls a "mediating artifact" (290)—for understanding the body and organizing the distributed work of dissection. That is, participants learn to make the body resemble atlas images—idealized, normative, and unchaotic. Dissection is the most remarkable lab event not merely because the experience is unique, but also because it represents an instance of embodied rhetorical action that develops a crucial aspect of anatomical vision, allowing participants to recognize and appreciate the cadaveric body's aesthetic dimension.

Dissection as Distributed and Enacted Cognition

During a prep-lab session for the dissection course, I observed the TA Ruth and an instructor dissect and tag a cadaver for an upcoming exam. "Make it the most beautiful they have ever seen," the instructor told her, suggesting she keep the major landmarks in place. Soon he walked over to another cadaver and began dissecting. As I watched, he used scissors to clean up the structure he wanted to tag by removing excess tissue around it. "I'm making a teaching example here," he said aloud to Ruth and me without looking up from his work. "I'm making the most beautiful iliac psoas they have ever seen." At this, Ruth came closer to have a look. "Making?" she questioned with a smile and the gesture of air quotes. "Well," he smiled, "I'm revealing the most beautiful iliac psoas they have ever seen." Obviously, the instructor does not make the iliac psoas but, as he points out, reveals it—yet this work of revealing inevitably transforms the cadaver into an idealized display, a paradigmatic multimodal illustration that students view to identify structures. The more *Netter*-like the dissection, the more students will be able to recognize it as demonstrating a particular structure. During the prep labs for the prosection course, the instructor guided TAs in this type of making. "So it would be nice if, when students see the ligament, it looked exactly like that." Pointing his finger to an image in *Grant's Dissector*, he spoke to the TAs, Martha and Nina, who were dissecting ligaments. "So we should get rid of all of this," Nina asked, pointing to tissue and structures close to the ligament. "Yes," he affirmed; they should "netterize" the dissection to make it resemble the image so that students will not question what they see. These TAs worked to reveal the structure and make it into a *Netter*-like, or netterized, example, one with (seemingly) unquestionable resemblance to *Netter* and other naturalistic displays. Here, making the cadaver beautiful means making it more presentational and connects students and TAs to the larger history of anatomical image making.

Historians of anatomy have described the ways anatomical images and anatomical dissection mirror each other. Many famous anatomical illustrations, like Vesalius's *Fabrica* (1543) and Gray's *Anatomy, Descriptive and Surgical* (1858), were based on actual dissected bodies (O'Malley 1964; Richardson 2008). Anatomists and students then used these images to guide the complicated and at times grisly work of cutting into other bodies. Even today, students, TAs, and instructors use atlas images to guide their dissections, cutting the body to match as much as possible the naturalistic displays on the page. Focusing specifically on surgical practices, Hirschauer (1991) finds anatomical atlas images to be "a normative picture" (311) of the body used to "document products of dissecting labor," thus offering an "idealized account of what has been done" (310). These displays shape not only the process of dissection but also the product that dissection creates. For those trained in the gross lab, naturalistic displays become the screen through which they see the physical body, and the process of dissection becomes the means by which they make that physical body into an object at once scientific (in the sense of anatomical) and beautiful (in the sense of conforming to the atlas's idealized aesthetic).

Dissection makes cadavers into typical and characteristic examples of the anatomical body; yet this making of one conceptual body occurs through the unmaking of a real cadaveric one. Hirschauer (1991) has suggested that what to a "surgeon" (or even an anatomist) might seem like "making anatomy" would to anyone else seem more like disfiguring a body (301). To create clear anatomical bodies, participants carefully and thoughtfully engage this process of making through unmaking. For example, during the dissection course's morning lecture on the flexor forearm and the palmar hand (or palm-side of the hand), the presenting instructor reminded the class that cadavers had "two sides"—left and right. As such, he informed them to consider dissecting one side superficially, while "sacrificing the structures" on the other side by dissecting deeper. The instructors in both courses often describe deep dissections using the language of archaeology; students and TAs should "excavate" an area of the body layer by layer, "digging" down to access the strata beneath the surface. In so doing, they not only disturb the dig site, they remove its superficial surfaces all together.

Judging what to "sacrifice" can be a difficult decision as one team in the dissection course illustrates. All four team members at one cadaver table engaged in a discussion of where to cut the cadaver. While the student Deena, with scissors in hand, tried to figure out how to proceed, her teammate, Sean, suggested she cut "here," pointing to an area on the body with his finger. He directed the team, specifically Deena, to a cut that would allow them to place the skin back over the body when they were done, to keep the body moist. Looking over at the same structures on a female cadaver behind them, Deena replied, "hers look different." By comparing her team's cadaver to one at another table, Deena confirmed that her suggestion was perhaps the better one. At this point, Michelle, another teammate, directed them to *Netter* so that they might double-check the structure's look and location. After leaning into *Netter*, Deena replied, "Oh, O.K.," as if the image was particularly illuminating. Deena and Michelle smiled at each other and looked at the book, before making a face as if bracing themselves and anticipating the worst.

This scene provides a representative example of the embodied, deliberative actions involved in dissection. Before team members can decide the optimal cut, they must determine the specific location on the body; they do this by turning to the multimodal displays as roadmaps for their destination and the route they should take. They cross-reference these choices with the decisions their peers have made by looking over at other students' dissections. This work requires trained vision to recognize the anatomical body and manual skills to make it visible. The collaborative teams help participants make these decisions and perform these operations. Still, according to nearly everyone I interviewed, dissection involves a great deal of trial and error. Participants must find and identity structures in the cadaver's frequently chaotic, always individualized 3-D space, and they learn to do so through an adroit interaction between seeing and touching. Though they are surrounded by a host of multimodal displays, verbal descriptions, and additional examples in their peers' dissections, participants must read the anatomical body on and into these displays and objects. They do this by incorporating the affordances of these objects into their bodily gestures, movements, and decisions so that they might use these objects to see and make the anatomical body—thus extending or distributing their bodily capacities by way of these objects.

Goodwin (1994) illustrates a similar phenomenon, a phenomenon he famously terms "professional vision," by investigating two disciplinary sites where participants learn to see in ways determined by their disciplinary activities and to demonstrate to others what they see through the tools of their profession. Goodwin emphasizes three discursive practices integral to the formation of professional vision: (1) coding, or the transformation of phenomena into "objects of knowledge"; (2) highlighting, which marks and makes salient phenomena; and (3) "producing and articulating material representation," which might include visual displays or informational graphics, from transcripts of conversation, to video, to charts and images (606–7). That is, participants develop trained vision through the strategic, and I would add habitual, deployment of some coding scheme or, in Goodwin's words, a "graphic representation," that "organizes" our perception of the objects and actions of an environment (609). In the development of technical expertise, a participant engaged in situated practice filters her perception through coding schemes provided by her disciplinary or professional community (616). According to Goodwin, for the field archaeologist, this filtering of perception involves using a Munsell color chart to see scientifically valid data in dirt; for the police expert testifying in court, filtering can mean interpreting a videotape of the Rodney King beating to demonstrate evidence of a suspect's body resisting arrest. Thus trained vision—learning to see, think, and even embody knowledge as a practitioner—involves learning to use tools as a member of that group. However, as Goodwin's analysis of the King trial makes clear, trained vision is never value-free or ideologically neutral.

As a skilled capacity, trained vision also requires what Ihde (2002) terms "the full-body perceivability" of our actions. By referencing a physical anthropologist at a dig site, Ihde illustrates how "informed vision" (as he terms it) is deployed along with our "interaction" with the environment (39). Through hands-on material work, a newcomer learns to see as a scientist or expert as he or she comes to embody the practices of that domain. This hands-on material work thus allows the newcomer to see and make material a particular type of knowledge while being shaped by that knowledge. In this social model, as with cognitive enaction, perception and cognition are formed through our social interactions with the objects and people of our profession. Hutchins (1995) describes these social interactions between people and objects as forms of distributed cognition. In his ethnography of the physical, social, and cognitive practices involved in ship navigation, Hutchins conceives of distributed cognition as a way of understanding the "social distribution of cognitive labor" experienced by groups of people organized to perform tasks (228).

For Hutchins (2006) and others (Griere and Moffatt 2003; Winsor 2003), distributed cognition is not merely a form of cognition that occurs in groups. Rather, it is "a perspective on cognition" (Hutchins 2006, 377); it is based on the premise that cognitive processes are "always distributed in some way" (376). Cognitive tasks are made possible by bodies and objects engaged in activity—whether a group of people together or an individual alone working with physical tools and texts. This sharing or distribution allows us, as groups and as individuals, to perform tasks in ways we otherwise could not without this division (Hutchins 1995, 2006). The tools and texts that allow us to engage in these tasks are what Hutchins describes as "mediating artifacts," or "structural elements" that "are brought into coordination in the performance of some task" (1995, 290). These artifacts can take many material and immaterial forms, such as a "written procedure," a tool, an "arithmetic procedure," a "language," and even "mental modes" (290). Through these mediating artifacts, whether objects, bodies, or discourses, human cognition is distributed across "brains, bodies, and a culturally constituted world" (Hutchins 2005, 376).

More recently, Hutchins (2010) characterizes distributed cognition in the language of enacted mind as opposed to the computational language of his earlier work (1995). Rather than view cognition as a brain-bound or even body-bound process that gets distributed across mediating artifacts, Hutchins contends that cognitive systems, as both embodied and enacted (2010, 428), "transcend the boundaries of individual bodies" (426). By way of mediating artifacts (objects, discourse, documents, and displays), we enact the objects involved in our situated practices through the creation of "enacted representations" that are "dynamic," "multimodal," and contingent upon our sensorimotor capacities (434). In other words, the embodied and enacted mind does not simply use objects to complete cognitive tasks; the sensorimotor capacities and activities of the body enact cognitive and perceptual experience by coupling with mediating artifacts, which are themselves shaped by these enactive processes. Though perhaps a subtle distinction, this enactive approach to distributed cognition, I contend, allows Hutchins (2010) to better conceive how "courses of action" can through habituated practices "become trains of thought" (445). The enactive approach, as I demonstrate, also sheds light on activities of dissection by articulating how participants develop an anatomical-aesthetic sensibility through their repeated use of mediating artifacts. Specifically, by using *Netter* as the model and guide for the distributed labor of dissection, participants netterize cadavers and enact an aesthetic orientation that trains their thoughts about the anatomical body. These dual forms of enactment—making a body and making a perspective on that body—develop through dissection's dual processes of revealing to learn and learning to reveal, both of which rely on enacted representations.

Learning to Reveal and Revealing to Learn

To successfully dissect is to mutually develop and deploy a trained vision that entails embodied interactions with the tools, instruments, and objects required to convert a dead body into an anatomical specimen. Dissection requires a bodily engagement with the objects that shape the dissector's living body and the (supposedly) docile cadaveric ones, binding the bodies of both participants. In her study of surgical simulations, Prentice (2013) describes how the embodied practices involved in surgical dissection shape the one who cuts along with the one who is cut (100). During dissection, both bodies, the living and the dead, are mutually constructed through what Prentice terms "mutual articulation" (228), a term that describes the ways the dissector's body and the anatomical body "come into being together" through the physical, material, and conceptual practices of cutting and being cut (16). To dissect is to literally extend oneself into the bodily flesh of another (Hirschauer 1991, 299). In so doing, participants develop visual and haptic awareness as well as manual skill. As Allen, a TA in the dissection course, jokingly told a group of students: "See, feel, be the fascia."

To develop this visual and haptic awareness, participants must act on the affordances of cadavers, tools, and displays to make a particular structure viewable. They learn to view and use tools and instruments as a way of maneuvering through and interacting with cadaveric flesh. Such maneuvers and interactions require knowing anatomy and recognizing how a particular cut provides opportunities for viewing the body. In Hirschauer's (1991) words, "one must see to cut, and one cuts to see more" (200). Participants perform this embodied exploration by making the cadaver an illustrative example of anatomical features. Though students are not evaluated on their dissection technique, the ability to deftly maneuver through the body and comfortably and carefully manipulate structures serves a real purpose. If dissection is a process of learning the anatomical body (the object of study) and creating an example of that body (a multimodal object), students benefit from a greater ability to transform delicate interiority without destroying neighboring structures they might later need. Because students learn through descriptive and relational evidence, they cannot cut potential landmarks needed to make sense of a structure in question. Medical and dental students learn the anatomical body not only through the process of dissection, of finding structures for themselves; they also learn by way of the product of dissection— a cadaver made to reveal certain structures, a body exemplifying medical ways of seeing and knowing. Anatomical knowledge and bodily skill articulate each other, then, through the dissector's corporeal engagement with these anatomical objects, an engagement designed to make cadavers more anatomically demonstrative by making their structures clearly visible. Bodies and objects as well as knowledge and skill are, as Prentice (2013) explains, mutually articulated and enacted through anatomical dissection.

In the gross lab, this vision involves seeing in the physical body the abstract anatomical one and recognizing the dissected cadaver according to an aesthetic sensibility, one that views cadavers as beautiful or ugly objects depending on the dissector's skill and the flesh's compliance. This anatomical-aesthetic appreciation, made possible by the lab's anatomical displays, specifically *Netter*, operates as a mediating artifact through which participants come to view the body and by which they organize their distributed labor on the body. Dissection involves skillful use of and engagement with objects, tools, and instruments that participants deploy to make sense of the cadavers. Participants use the affordances of these objects (atlases, scissors, forceps) to cut into the body and, perhaps more importantly, learn to identify and reveal the structure in question. All of this they do by enacting a representation in and by way of cadaveric bodies.

Understanding how to begin and where to find structures are the first obstacles for beginning dissectors. As one medical student described to her TA during a dissection lab, "everything makes sense in the textbook, but not here." "Yeah," the female TA said, "but at some point it will." Pointing to the axilla (or torso) of the cadaver, the student seemed confused by the poor visual correspondence, or poor isometric compatibility, between *Netter* and the cadaver she needed to dissect. Here the student could not map *Netter* onto the body until she began dissecting away the superficial structures that stood between her and the structure in question. Before she could find "the cadaver one learns from *Netter*," she had to begin dissecting a surface that did not resemble *Netter*. The TA's words exemplify the educational processes of revealing to learn and learning to reveal.

Students in the dissection course and TAs in the prosection course describe this mutual relationship of revealing and learning as being "lost" in the body. First, they get lost in the sense of having lost their way and feeling confused about how, what, and where to cut. Second, they lose themselves by becoming intellectually absorbed in the process of dissecting. In the first case, the losing of one's way, participants rely on instructors, TAs, or students to step in and begin identification, for example by calling out structures as a wayfinding aid. This collaborative element of dissection, working in teams to find and identify structures, supplements the use of multimodal displays when the cadaver's physical body resists a simple rendering or when there is little resemblance between the body and the display. As the course instructor makes plain, both cases common:

> [The structure in the cadaveric body] often looks nothing like it does in *Netter*. *Netter* is not that helpful then. It is a nice conceptual idea, but once you get in there [in the cadaver] it looks nothing like it. So you have to rely on your anatomical knowledge or someone else in the room. (Kyle, instructor of dissection course)

Dissecting is a process of learning to reveal that simultaneously allows participants to create the anatomical body by learning to recognize it in the unruly flesh of the cadaver.

In the second meaning of getting "lost," participants are lost in the revealing to learn component of dissection, due to the intense concentration and deliberate actions dissecting requires. They need to find and identify structures in question and continue to dissect and, in a sense, perfect the cadaveric body to make those structures clear and obvious to anyone who views them. They become intellectually absorbed in their work, lost in the body, because making a structure clear and visible requires minute, detailed operations. While at times tedious, this work can be, for all participants, enjoyable. They develop manual skills and anatomical knowledge as they cut and recognize what they see, in the process gaining greater appreciation of the body's complexity and the elegance of its physiology and relational evidence. The work of making and making clear the anatomical body for oneself connects gross anatomy students and TAs to the centuries-old practices of anatomical display. Making cadavers into idealized 3-D displays offers them the pleasures of dissection and encourages in them an appreciation of the well-dissected body.

The Aesthetics of Anatomical Dissection

Dissection is a process that makes anatomical bodies out of cadaveric one, understood as beautiful according to the trained vision of participants. Though not attractive in a conventional sense, participants describe dissected bodies and the processes of learning to reveal and revealing to learn in aesthetic terms. This anatomical-aesthetic sensibility is a consequence of their budding anatomical vision, the use of *Netter* as a model, the finished product, the process of discovery, and the experience of working closely on the body. As a product, the beautiful cadaver is made *Netter*-like through a careful interworking of looking, touching, and cutting. The process of making, which creates the netterized body, facilitates a kind of pleasure rooted in the revelation—more specifically, the enaction—of a body participants deem awe-inspiring, through a process they describe as gratifying.

Making Netterized Structures

Participants in the gross lab characterize the well-dissected body as "beautiful" when it is "clean," "clear," and "lovely." One instructor of the dissection course explained to TAs that the best way to create good prosections was to "make them beautiful; make them clear." Then he added: "But don't just make them clear, make them clean, make them *Netter*-like." The expertly dissected body, or prosection, is the one that most resembles the idealized image presented in *Netter*. The beautiful or "clear" prosection, then, is simultaneously characteristic and typical, like naturalistic displays found in a variety of atlases. In the gross lab, the beautiful body is one devoid of obfuscating fat and fascia, absent distracting and confusing structures. At the same time, this body demonstrates landmarks and relations necessary to identify structures in question, thus allowing dissectors (students, TAs, and instructors) to exhibit anatomical knowledge and manual skill. The dissector must know exactly what to remove, what to keep, and what to netterize to make structures in question obvious and emblematic, didactic and authentic.

In particular, as mediating artifacts, visual displays act as the coding schemes participants use to transform the unruly cadaver into a beautiful anatomical specimen. As in-process instructional documents, visual displays provide clues on how to proceed. They also serve as end-product illustrations, showing what the results of the work should resemble. The normative, naturalistic displays (like atlas plates or photographs) are particularly important tools for guiding the dissections of more advanced dissectors (like TAs). During one prep lab for the dissection course, one TA, Ruth, brought in laminated photographs of cadavers. "This is how our dissections are supposed to look at the end of the day," she told her fellow TA Brian, who then asked to see the photograph of the suboccipital triangle because he wanted to use it to guide his dissection. In particular, students view *Netter* plates, though at times cluttered and deceiving, as ideal guides for dissection:

> I try to look over my *Netter* in order to get a picture, a visual picture. Because I find that if I don't have any idea what it is supposed to look like when I go in there, you know, you tend to kind of ruin things. And if you know where to look for things, you know different relationships to keep an eye out for them. Because it seems like if you can preserve them, you will get and see a lot more important things. (Carlotta, medical student)

Netter provides a necessary image that operates as a map for finding structures in question without destroying structures they will need later. In the undergraduate prep lab, the instructor requests that TAs confirm their dissection by returning to *Netter* periodically as they work. "You can see where you're at in this stage," he told one pair of TAs dissecting for the exam. By pointing at a plate in *Netter*, he had them confirm their work even though they had not yet reached the structures in question. Here he turns to the book because these structures would not be visible until TAs had completed a great deal more dissecting.

The netterized body is also valuable for tagging structures that students must identify for formal exams. As I mentioned in chapter 4, a teaching tag, or "beautiful tag," as it is called, provides a view of the structure in question that is obvious to the trained participant. Participants create a teaching tag by removing all excess and distracting structures that are unnecessary for identification. By calling it a teaching tag, participants imply that the tag can be used as a teaching example, one student dissectors should strive to emulate, and an example that teaches the anatomical structure in question. During the set up for an exam in the dissection course, one instructor called another over to the light box to view an x-ray tagged with a green arrow. "Does that not bring tears to your eyes?" he jokingly asked his colleague, pointing proudly at the x-ray. These metaphorical tears were brought to the eyes by the emblematic and expertly rendered tag, which made identification possible without obscuring necessary landmarks. As I discussed previously, creating teaching tags is complicated in part by the cadaveric body's tendency to resist these rendering practices.

The Pleasures of Making

Part of dissecting a cadaver is uncovering its anatomical complexity, which students and TAs often experience as a sense of awe and wonder. They might describe a structure as "awesome" when it is not merely *Netter*-like, but also illustrative of some fundamental physiological relationship or surprising in its look or texture. In the dissection course, one team praised their female cadaver's "awesome" rectus abdominus muscles, specifically the thickness of her muscles compared to the thinner muscles on cadavers at other tables. The team could easily find and view these "awesome" muscles as well as appreciate their relationship to neighboring muscles in the abdominal region, thus gaining a better view of physiological relationships. During another dissection session in this same course, I watched one student manipulate the greater omentum of his team's cadaver (a large fold of tissue that hangs down from the stomach and over the intestines). He repeatedly ran his gloved hands over its ruffled surface the way a tailor might touch fabric. In fact, he compared the look and feel of these membranous folds to chenille, urging his teammates to feel for themselves. Though they debated the structure's resemblance to fine fabric, each was surprised by the tissue's look and feel that seemed, as one student put it, "manufactured and not like something natural." Later, an instructor called everyone's attention to that cadaver's "classic, spectacular greater omentum" and urged everyone to "see and feel" before leaving lab.

Participants often invoke a cadaver's beauty in connection with their perception of the body as a beautiful machine. One dental student, Gordon, describes a feeling of awe that directly references a mechanical metaphor:

> Just to sit there, sometimes you are just totally into what you're doing, [listening to the prosection demonstration] and you kind of just view it as a science experiment. And other times you step back and you realize, wow, just what a phenomenal thing it is, just like the greatest machine, if you really look at how things are put together.

For Gordon, the conception of the body as a science experiment, emphasized by prosection demonstrations on the location and function of structures, gives way to an idea of the body as an impressive technology. These two related conceptions of the body—as science and as wondrous machine—are one and the same for many students in that both seem to result from the aesthetic appreciation of complexity. In discussing the benefits of working on cadavers, Miles, a dental student, likens the body's complexity to an automobile: "Like, you know, all the Honda engineers could work a century, and they still would not be able to design everything in the body." Gordon's language, like Miles's language, is reminiscent of what was once understood as the philosophical or moral aspects of anatomy expressed in the Roman physician Galen's second-century CE text *De usu partium corporis humani* (in print since 1528). This primarily philosophical treatise stressed the wisdom of God's design as made evident in the body. In today's gross labs (as it was hundreds of years ago), the science of the body is part of what makes it beautiful.

Connected to this is an appreciation of the cadaveric and living body's fragility. When I asked participants about their favorite anatomical structures, or the most memorable ones they encountered, most students, in both courses, mentioned the hair-like fibers of facial and cranial nerves:

> The one thing that will always stick with me are the cranial nerves because of how delicate they are. You know, you're looking at the structures that are so delicate that if you were to hit them with just a little bit of force they will break. Yet, you run into things all the time, and you hit your head all the time. And you wonder how these things don't just break, or you wonder how these nerves, and how the structures are not damaged. (Barbara, prosection course TA)

Barbara is amazed by the delicate, yet durable cranial nerves, which are difficult to dissect without destroying. In the undergraduate prep lab, these structures are dissected usually by the more advanced TAs, who work slowly to remove concealing fat and fascia.

The body's delicate complexity makes dissecting certain structures, like cranial nerves, an intense and rewarding process because careful and skillful dissection makes this complexity viewable. According to an instructor in the dissection course, dissecting the arm is "very gratifying" because dissectors can easily find, identity, and netterize key muscles, arteries, and veins. Body-buddy teams can be noticeably excited by the work of dissection, specifically after spending much time and effort locating structures that either resemble *Netter* or visually suggest a structural relationship. One team in the dissection course once called me over to their table because they "found something cool." As I moved in for a closer look, one student pointed to a space formed by four muscles of the arm: the teres major, teres minor, the long head of the triceps brachii, and the neck of the humerus. "The quadrangular space," he told me, as if presenting it for my inspection. These muscles together form an easily recognizable quadrilateral shape of sharp angles. And though this anatomical space is deep inside the arm, participants can easily netterize it so that it resembles an atlas illustration.

Dissection, making visible the anatomical body in cadaveric flesh, is a labor-intensive experience marked by a kind of emotional intensity. Students and TAs are absorbed physically in the process of dissection, immersing their hands and sometimes arms in the body's viscera. But dissection is intellectually and emotionally enthralling as well. One day in the prep lab of the dissection course, I stood near the sink, taking notes, when a TA, Betsy, walked over to get a plastic apron. "It is getting intense," she told me, "because it is so exciting, following everything through." After returning to her cadaver to continue dissecting, she called over to another TA, Frank, to tell him she has found the pudendal nerve inside the cavity of the pelvic floor. She had found it, she explained, by locating the internal iliac artery, and she wanted "to track it" all the way in the body, to be able to show that to students. "Oh my God, I'm so excited. I'm tearing up." Though she was joking, she did appear to be tearing up in exhilaration, even joy.

Though not always to the point of tears, participants do express visible excitement when locating a hard-to-find structure in question. Often they call out to TAs and students to show off their work on the cadaver. One female medical student, Carla, could at times even be seen talking to more elusive structures, like the day one of her teammates teased her for saying "hello" to a deep muscle she had been working to uncover for some time. She smiled at this good-natured ribbing from her peers and explained to them (and me) that she greeted it because, in her words, "When I found it, I knew exactly what it was." Dissection, as an enactive process of revealing to learn by learning to reveal, can be a rewarding and pleasurable experience. This making and making visible of an idealized anatomical body, however, is not always possible.

Cadaveric Resistance

By moving the cadaveric object from chaos to order, from unruly flesh to beautiful display, dissection transforms the dead body into an exemplary presentation of the anatomical body, or more specifically, of structures participants must find, identify, and learn. As such, dissection often involves getting lost in the chaos of the body in order to reach the final goal of ordering. This is not to say that cadavers remain docile objects that comply with these rendering practices. Cadavers can be recalcitrant and completely resist this act of making through the abjection of the dead body and the presence of anatomical anomalies. Both of these complicate the pleasure of dissection and the beauty of the dissected body. These instances of recalcitrance, however, are inevitably part of the process of learning to reveal, through which the anatomical body is made and with which participants must learn to contend.

Abject Bodies

Kristeva (1982) has theorized abjection as represented in and by the body's secretions and open orifices, namely "jettisoned" objects like excrement, blood, and pus that are "radically excluded" from the body (2). More fundamentally, the abject, which lacks a concrete "definable object," is that which "cannot be assimilated" by the body and more importantly the "I" of personhood (1). This challenge to the I, to the ego (in her largely psychoanalytic model), is the quality all abjections share. The human corpse or cadaver is "the utmost of abjection," in that it represents and in a sense *is* the destruction of the I (4). The cadaver undermines our confidence in the security of living by underscoring life as "fragile and fallacious" (3). Further, the cadaver makes visible what the living human must "thrust aside in order to live" (3). "It is death infecting life," Kristeva tells us (4). After all, abjection, which finds a vivid and durable instantiation in the unclean or dead body, ultimately originates in that which "disturbs identity, system, [and] order" (4). Schwenger (2000), extending Kristeva's notion, emphasizes the dead human body as a kind of "uneasy frontier" that is both subject and object, yet neither (400).

In the gross lab, the abject body is visible in the physical excesses and recalcitrance of the cadaver, which is disciplined and even sacralized through dissection. In the gross lab, the abject body emerges when the cadaver resists the ordering practices (namely dissection) that make dead flesh visible as anatomical systems. First, the embalmed body is stiff and not always easily manipulatable. Early in the dissection course, students learn to contend with the cadaver's rigidity as they struggle to move, for example, a dangling arm that seems to block their way. Getting close to cadavers sometimes means, as one female student put it, "figuring out how to get over" an arm or leg that seems to obstruct your work. She "gets over the arm" usually by stepping inside the space it creates—at times standing close to a cadaver whose arm is wrapped around her waist.

Second, the cadaveric body is abject in relation to its wetness and the persistence of bodily fluids. Though embalming removes the body's blood and infuses it with liquid fixative agents, participants keep the cadaver damp to prevent the tissue from drying. Each day before the cadaver is lowered into its tank, a student or TA sprays the body down with a mixture of water and acetone. Along with the oil from the body's adipose tissue, this liquid, which pools on and in the concave surfaces and interiors of the body, can make dissecting messy. Once in the dissection course, I watched a male student flinch and step back abruptly, wiping his face with the arm of his sleeve. "Are you OK?" a second male asked, grimacing. "Yeah," responded the first, "but I got some of the fat in my face." He seemed to be smiling and grimacing at the same time as he walked to the sink; his teammate nodded and smiled with empathy.

Scenes like this are common during dissections, and result in as many smiles (of commiseration and empathy) as hurt feelings (by those frustrated with what they view as a teammate's carelessness). Also, as a human body, the cadaver still contains traces of fluids, secretions, and wastes necessary for and byproducts of biological function. There might remain some excrement in the intestines; an undetected blood clot might leave small amounts of blood. These traces of human physiology cannot and perhaps should not always be expunged from the cadaveric body because they may provide information that aids in the identification of structures and relationships. One team's work in the dissection course was delayed for several minutes because the team's cadaver appeared to be bleeding. Traces of a translucent red fluid had pooled beneath some sections of the body. By following the path of the flow to its origin, the team physically located and verbally identified the artery that seemed to cause the leak.

One goal of skillful dissection is to reduce the body's abjection by eliminating unnecessary or unsightly structures that might detract from the utility and aesthetics of a beautifully netterized cadaver. In a prep lab for the prosection course, the instructor advised a pair of TAs to remove excess skin so that the prosection would not, in his words, "gross out the students." Specifically he called their attention to one flap of skin they should leave on the body (because it could be laid back in place) and another flap they should remove (because it obscured the area and served no purpose). It would be impossible and unhelpful to completely remove or eliminate abjection, however. Dissecting a human body, cutting the face, the hands, genitalia, or removing limbs: these tasks inevitably make visible the abject that we are, the abject that our body quite naturally is. Some dissections do seem to showcase a potentially unsettling view of the body as both cadaver and former person.

When I asked participants about their most difficult or challenging dissection, most recounted dissections they found emotionally unsettling, as opposed to physically difficult. They overwhelmingly described dissecting faces, hands, or genitalia as difficult because it involved cutting into complex, minuscule structures that conveyed a personal humanity. We recognize each other, seek out and speak to one another's face, which we conventionally understand to be that person through a synecdochic formation that conflates the face with the self. Levinas (1985) describes our relationship with, even "access to," the face of another as "straightaway ethical" (85). Our responsibility for others forms our "authentic relationship" to them (88) and is witnessed in and witnessed as the "essential poverty" of the face (86). Moreover, we use our hands to interact with and even form the world around us. Participants spend so much time learning the anatomical body by touching and moving it with their own hands, but in the dissection lab, those same students have to dissect the hands of another. Finally, due to prohibitions that require us to conceal our genitals, as well as the ways we understand our sexual organs, pleasures, and acts as representing something fundamental about individual and group identity, our genitals seem to symbolize something unique about us as well.

Dissecting these intimate structures means revealing the anatomical body we are by going beneath the individual self we seem to be. As one undergraduate TA articulated it, "I am inside this person's face and then I'm thinking about my face." Dissection, like abjection, confronts participants with a kind of I-yet-not-I that the corpse seems to signify. Yet this abject body is unavoidable and, in a sense, natural; it is part of the biological basis on which the human body functions. The cadaver seems to remind us that we are our physical bodies, and yet we are not only our physical bodies. This material physicality of human flesh is unruly and confusing until a seemingly elegant order is placed upon it, revealing in its excesses and differences through an enactive representation that makes it somehow beautiful.

Anomalous Structures

The beauty of the unruly body made orderly finds its best contradiction in the anatomical anomaly, which participants view as simultaneously natural and unnatural. Anomalous structures are natural in the sense that they commonly occur in the body and not as a result of human intervention; they are unnatural in that they are nonnormative or atypical. The cadavers of the gross lab, like all bodies, contain anomalous structures that offer unexpected variation and exceptional deviation from the normative anatomical body. Participants in both courses exhibited to each other, the instructors, and even me any noticeable anomaly. In the words of students and TAs, there were "weird arteries" that were larger than normal, and "huge ones that usually aren't there." There were cases of "too many arteries" and, on one cadaver, "alien arteries" that were too many, too large, and oddly located. These anomalies, though confusing, test a participant's anatomical knowledge, observational skills, and dissection technique by requiring her to identify even anomalous structures in question and then match those with their nonanomalous examples in other bodies. In the prosection course, one TA, Doug, reassured a student simply by acknowledging the presence of anomalies. "There are a lot of weird fascial planes on this body," he told one female student, who smiled and replied, "I'm so glad you said that." With a look of relief, she thanked him for confirming her suspicions.

Some anomalies can be startling, like normative structures occurring in unexpected locations. Some anomalies are deemed so unusual they take on an almost celebrity status. In the undergraduate prep lab, the mysterious tumors of one cadaver invited a closer look from nearly every TA. When anyone dissected a structure on that cadaver, they would often take a moment to inspect with eyes and scalpel the unknown growths located in the cadaver's abdomen. The TAs talked about and repeatedly examined them; some even turned to textbooks in the hope of diagnosing them. In an interview, TA Doug even admitted "there was an obsession with those tumors this semester." Anomalies represent components of the anatomical body that are deemed simultaneously natural (biologically occurring) and unnatural (nonnormative). Deviating from the normative body, and thus unable to be visually ordered and made beautiful, anomalies are either removed to create a more netterized corporeal display or left in place to illustrate the importance of recognizing variation.

Some anomalous structures eliminate altogether the cadaver's ability to enact the anatomical body. These anomalies, specifically the physical consequences of disease or surgery, can impede the cadaver's viewability (one of its major affordances) by interrupting the ordering practices of dissection. The presence of excessive tumors, the ravages of cancer, the damage caused by stroke: these can prevent the making of aestheticized anatomical bodies because they complicate a participant's dissection. While dissecting their cadaver's brain, one team of medical students discovered evidence of a stroke, namely a severely blocked vessel and the resulting damage to the brain. Because a stroke was not listed on the cadaver's medical history sheet, the team stopped working and called over a professor, who confirmed their hunch. Though dying of another ailment, this donor had indeed suffered a stroke that left a considerable mark. Unfortunately, the instructor had to remove the brain from the cadaver because the obstructed vessel might have hampered the circulation of embalming liquids needed to preserve the tissue. Disease had, in effect, made this precise structure unviewable. When one of the team members spoke to me about it later, she expressed visible sadness for her cadaver and her group. The traces left by this stroke reminded the students of "where these bodies come from," namely "people who died of something" (Kayla, medical student). Yet the stroke also meant her team would not have an anatomical specimen to dissect and study. Though her sense of sadness over her team's misfortune might seem unfeeling, this dual sense of sadness (for another and oneself) is oddly reminiscent of the grief we express whenever anyone we care for dies; we mourn both their passing and our loss.

The body wrecked by disease represents the limits of the anatomical making that characterizes cadaveric dissection. These particular bodies refuse the aestheticizing practices—the hard work of learning to reveal and the gratifying pleasures of revealing to learn—because they cannot be made *Netter*-like. These bodies will not be clean, clear, and beautiful. This recalcitrance does not render them useless, however. During one memorable afternoon in the dissection lab, I witnessed two teams huddled around one cadaver. I walked closer to find them looking into the abdomen of a female body. Before I could get a good look inside, a female student at the table said to me, with a somber tone of voice, "cancer." One male student, Tommy, used a scalpel to lightly scrap the surface of the large cancerous gray-green mass that seemed to engulf the abdominal region. "So is that—?" I started to ask, and before I could finish, he replied, "It has just eaten away at this cadaver." Suddenly, Tommy was visibly grossed out. He began blinking heavily, and nearly gagged. A female student, Carla, reflected, "This [cancer] has to be more painful that I can imagine." As the others looked on, they attempted to identify the structures that seemed untouched or at least still recognizable. Carla seemed to search for ways to describe what she was witnessing: "It's creepy, and sad, and . . ." She stopped and looked over at an approaching male student, who reacted in astonishment as he looked in: "Holy crap." Another female student responded, "There is very little left." Soon the TA, Allen, walked up to check out the situation, no doubt wondering why so many students had crowded around one table. He looked into the body, frowned, and confirmed that it was in fact cancer. He added, "It makes it difficult, that is for sure."

The unexpected presence of cancer not only stopped the team's dissection; it cast a shadow over those close enough to see the disease's devastating effects, which had altered most of the structures in the body's abdomen. As they stopped to inspect, students experienced a number of reactions: they were puzzled by the anomalous mass, shocked by its visible effects, repulsed by the sight of the altered tissue, and saddened by the thought of this donor's experience with disease. The TA's comment about the difficulty of the situation, I contend, is revealingly ambiguous. On one level, Allen is referring to the sadness of finding cancer in the body of a person who experienced great pain and a host of physical complications that made her final years of life difficult. This cadaver's cancer palpably brought the patient-body into focus, by exposing the physical effects of disease and providing participants a chance to consider what is termed "the hidden curriculum" of the lab—the presence of death, dying, and inevitable mortality (Hafferty and Franks 1994). Yet, on another level, the TA's comment also alludes to the difficulty of identifying and learning anatomy in a cadaver transformed by a disease that obscured and even destroyed anatomical structures.

Two days later, after hearing a prosection presentation at another table, one of this cancerous cadaver's body buddies asked his teammate, "Did you see all the things you can see in that body over there?" The cancerous anomalies of their cadaver made its abdominal region anatomically unviewable. Later, during an interview with one of this same cadaver's body buddies, I learned that the team began to feel uncomfortable with the way students came over to inspect the cadaver. Though she admitted her peers needed to take a look and understand the effects of cancer, she felt almost protective of her cadaver. At the risk of sounding unfeeling, I want to suggest that this woman's cancer also provided an opportunity the noncancerous body does not. It underscored the artificiality of the netterized body and the naturalness of the unruly, abject one. The cadaver's resistance inevitably emphasizes the usefulness of *Netter*'s idealized bodies and the skillful prosections performed by TAs and instructors. This resistance also inevitably fosters an appreciation of expert dissections made possible by the skilled technique of a more advanced anatomist.

The Aesthetics and Spectacle of *Body Worlds*

Von Hagens's *Body Worlds* physically presents, in the bodies of plastinated cadavers, a significant example of the distributed cognition and anatomical-aesthetic sensibility involved in advanced dissections, both of which make anatomically beautiful the abject, unruly body. During my fieldwork, Von Hagens's exhibition was on view at the local science museum, and many anatomy students, TAs, and instructors visited the exhibit. Every participant I interviewed who had seen the show praised it for its presentation of "perfect" bodies that "beautifully displayed" the "complexity of anatomy" (all phrases taken from interviews). Similarly, many medical and dental students appreciated how *Body Worlds* seemed to address what they understood as one pleasure of anatomy education, specifically an appreciation of the awesome complexity and intricacy of structures, made visible and beautiful through skillful dissection. Students and TAs marveled at the unmatched dissection techniques of Von Hagens and his team. One medical student was "really impressed by the level of detail, and how they were able to keep everything so intact." Several participants admitted to taking their family and friends to the exhibit to "kind of let them see what we do in the labs," and in part to encourage them "to appreciate their bodies more." This is not to say that students, TAs, and instructors did not question some aspects of the exhibit, such as the cadavers' often elaborate poses, the isolation of the pregnant cadaver and the fetuses, and the ethics of making a traveling show out of such a display. But as I illustrate in this final section, *Body Worlds* took on a greater significance to the participants of the lab because the exhibition instantiates the categorical complexity of the cadaveric body as gruesome and beautiful, scientific and artistic, a process of looking and a product of vision, nature and culture.

Since its early exhibitions in the late 1990s, Von Hagens's showcase of plastinated cadavers has been surrounded by controversy involving the ethics of displaying dead bodies, the exhibit's ideological tone, concerns over Von Hagens's supply of corpses, and the details of the informed consent process. The latter two concerns have largely been put to rest, thanks in part to an ethical review of *Body Worlds* completed by the California Science Center in 2005 (California Science Center). Van Dijck (2005) identifies what is perhaps "most unsettling" about Von Hagens's work, namely the ways in which it challenges the conventional "epistemological categories that guide us in making all kinds of ethical distinctions" (62). One such distinction is the one we make among scientific and artistic objects and ways of looking at those objects (Van Dijck 2005, 60). Analyzed with this distinction in mind, Von Hagens's cadavers help reveal the trained vision of anatomical practice. As I discussed earlier, scientific and artistic vision converged in the history of anatomical practices, images, and displays. *Body Worlds* quite self-consciously takes up this long anatomical tradition by incorporating seemingly artistic conventions and aesthetic concerns (Van Dijck 2005, 53). These conventions and concerns are apparent in his visual references to famous historical anatomical images, such as the cadaver modeled after an *écorché* (or self-flaying man) illustration from Valverde de Amusco's *Historia de la composicion del cuerpo humano* (1560). A museum visitor also witnesses Von Hagens's aesthetic and philosophical concerns in his decision to incorporate prints from Renaissance artists and ideas from enlightenment thinkers. In particular, the American exhibitions of *Body Worlds* seem, to Schulte-Sasse (2006), to be "a celebration of enlightenment [and] scientific progress" (374). The show exhibited near the location of my fieldwork, for example, presented cadavers arranged amid giant artistic reproductions of the great masters, such as Rembrandt's *The Anatomy Lesson of Dr. Nicolaes Tulp* (1632), as well as quotations on life, death, and humanity from Descartes and Kant. (Schulte-Sasse [2006] provides a vivid description of the same *Body Worlds* exhibition that many participants of my study visited more than once.)

As a result, the space seems to invite a kind of moral and philosophical reflection on the meaning of life and the nature of death (Moore and Brown 2007, 237). Von Hagens himself has described his exhibits as spaces "of enlightenment and contemplation, even of philosophical and religious self-recognition" (Institute for Plastination 2013). By rhetorically framing these showcases of bodies in this way, Von Hagens is borrowing from early modern notions of the moral anatomy lesson, in which dissectors and spectators used the public dissection theater as an opportunity to meditate on the wondrous handiwork of God (Guerrini 2006). What to some viewers might seem questionable—the way *Body Worlds* uses this philosophical orientation as a rhetorical frame—was to students and TAs an understandable and laudable connection. One dental student explicitly connected Von Hagens's cadavers to Vesalius's atlas, asserting that both served a "scientific and artistic function."

Still, the presence of stylistically and artistically posed cadavers has proven problematic to critics and students alike. Walter's (2004) study of audience reactions to *Body Worlds*, drawn from interviews, surveys, and comments from the guestbook and website, finds that most visitors either loved the cadavers' elaborate poses or were "disturbed" by them (469). By positioning cadavers to represent living bodies engaged in sports such as basketball, running, or archery, Von Hagens seems to invest these plastinated bodies with a "social identity" as human beings in action (469). Schulte-Sasse (2006) astutely attends to the more troubling "cultural implications" of the athletic poses, namely an "aesthetic" reminiscent of Nazi art and representation that often presented the human body in a "gloriously idealized form" (379). Though she makes clear that labeling "Von Hagens's aesthetic" as "fascistic" would be an "oversimplification," (379) her reading of these posed bodied illustrates "the overwhelming power of context" (374) and the troubling connotations of the normalizing impulse of anatomically beautiful bodies. This possible connotation was not lost on laboratory participants, one of whom even asked during our interview if "a German man should be posing bodies in that way." Though Von Hagens was born in Poland and raised in East Germany, this undergraduate's question references one criticism that viewers raised against these stylistic poses that made the exhibition seem, in the words of one dental student, "not completely educational and a bit ideological."

Yet the staging of bodies in athletic or active poses has been a part of anatomical demonstration for centuries. Vesalius's famous muscleman images of the *Fabrica* imply the anatomical function of the body's major muscles by demonstrating a flayed man walking in nature. The atlas makers Cheselden and Albinus not only staged cadaveric specimen, they (and their illustrators) created atlas images modeled on those stylized poses (Daston and Galison 2007). Even today, anatomy professors incorporate images of athletes or perform athletic poses to teach the structural and physiological relationships of muscles. During the lecture on the extensor forearm in the dissection course, the presenter included in his PowerPoint slides an image of Pete Sampras playing tennis. Using a photograph of the tennis star taken at the moment before his racket makes contact with the ball, the lecturer directed students' attention to the motion of Sampras's extended forearm and his dorsum hand (the back of his hand) clenching the racket. Followed by a *Netter* illustration labeling the muscles of the forearm, this image offered a memorable demonstration of these muscles in action.

Hirschauer (2006) and Schulte-Sasse (2006) might argue that deploying an image of an athletic body illustrates the normativity of the idealized anatomical body. After all, a visual display such as this references and enacts the ideal anatomical body by arguably conflating it with the idealized athletic one. Yet as I have mentioned, anatomy students in both courses appreciate, even perhaps advocate, the visual and kinesthetic affordances of a lean, physically fit body because it allows participants to see and feel the location and operation of muscles, veins, and arteries. From Vesalius to Von Hagens to contemporary anatomy professors, a trim, athletic frame is established inevitably as a pedagogical ideal because it allows the demonstration of structure–function relationships without the obscuring, abject unruliness of body fat. This is not to say that gross anatomy students and TAs fail to question this privileging of the athletic body, specifically in relation to *Body Worlds*. One dental student, Mandy, questioned why the plastinated male cadavers were frequently posed as athletes while the females were usually presented in motherly or supposedly feminine poses; she directly referenced the plastinated female cadaver arranged in a supine, almost seductive position. One undergraduate TA, Bill, even wondered what a general audience might make of the posed bodies; would they recognize them as indexing a long tradition of anatomical demonstration, as he seemed to do? Though ready to question and even challenge specific examples, most participants of my study read the posed cadavers of *Body Worlds* in light of that anatomical tradition of aestheticized and beautiful anatomical bodies.

As an anatomical and artistic spectacle, one seemingly designed to provide scientific knowledge and encourage philosophical reflection, *Body Worlds* questions many of the same supposedly durable categories disturbed by practices of anatomical dissection and display. One medical student's appraisal of the showcase emphasized the categorical contradictions made visible by the anatomical aesthetic: "Those bodies were incredibly dissected. You don't even think they're cadavers at all. It's just like perfect, picture-perfect bodies. It is amazing" (Will, medical student). Like his peers, he lauds the precise dissection techniques displayed by the flawlessly dissected bodies. This perfection, however, makes the bodies seem like something other than cadaveric specimen. Instead, the expertise of Von Hagens and his team has made these plastinated cadavers "picture perfect." Through this juxtaposition, Will implies that the cadaveric body cannot be perfected in the way these bodies have been.

Through dissection and the chemical processes of plastination, these bodies have been ordered, idealized, and even deabjectified to the point of resembling plastic mannequins more than human bodies. Again, plastination not only removes the body's fatty tissue, it infuses (or impregnates) the body with polymers at the cellular level, making the body human and more than human. Because they attain a level of perfection that the average cadaver and dissection

cannot match, Von Hagens's cadavers radically emphasize the anatomical aesthetics of dissection by bearing witness to the visual possibilities opened up
by the beautiful cadaver. They do so by inviting viewers to question our usual
notions of what it means to be human. For Kuppers (2007), these bodies, lacking any sense of abjection, offer "a realm of strange beauty" that "glistens and
beckons" (37) the viewer to observe and even envy: "the whole-body exhibits
are much more beautiful, much more classical, much more sure than the masses
wandering among them" (41). These unsure masses, of course, are the visitors
who move among these largely naked, largely idealized bodies, mostly staged in
athletic or active poses.

During my first visit to the exhibit, I was struck by the odd vitality of the
plastinates. Like exhibitionists in some macabre peepshow, the naked, almost
confrontational bodies were frozen in the performance of some physically challenging activity. As I strolled through in my own unruly flesh, gazing on their
often-statuesque forms, I saw that some bodies did, however, carry signs of disease. For example, cancerous lungs were displayed, as were glass tables containing
whole-body slices of obese cadavers. These nonidealized bodies are foils for the
more idealized one, throwing into relief the visible contrast between the trim
(and supposedly athletic) body, for example, and the obese (and supposedly non-
athletic) one.

Taken together, the bodies of *Body Worlds*, both the idealized and nonidealized
ones, inevitably taunt us. The presence of unhealthy bodies (which we fear we
are) seems to serve a public health function by calling attention to and encouraging physical fitness. One dental student, Naomi, claimed the exhibit made her
"appreciate life more." Another, Megan, claimed the exhibit encouraged her "to
be healthy and take care of" her own body. *Body Worlds* does this by highlighting not only the naturalness and inevitability of death but the work involved in
producing and maintaining life. The intricacy of human anatomy, the interde-
pendence of anatomical processes, the intentional human activity necessary for
health: these aspects of moral anatomy are silently referenced in the dramatically
posed dead on display, allowing audiences to experience and perhaps appreciate
the aesthetic orientation by which participants come to see the cadaveric body
as ideal. Or in the words of one medical student (David), "It wasn't just solely
an exhibition of human anatomy. It was an attempt to portray the artistic value
of anatomy." *Body Worlds*, I contend, is unsettling because it showcases an aesthetic that we normally encounter only by experiencing the gross anatomy lab
and developing the trained vision of anatomy, specifically dissection, that makes
this aesthetic possible in the first place.

Even with their trained vision developed through embodied (and embedded)
socialization, not all anatomy lab participants completely shared Von Hagens's
perspective. The students, TAs, and instructors I interviewed almost unani-
mously derided one stylized plastination. Staged to suggest a body caught in the
act of running, "The Runner" is a plastinate whose major muscles have been cut

at what appears to be the point of insertion, made to extend out into the air or dangle to the side (see figure 5.1).

Participants assailed this cadaver because, in the words of some, it was "ugly," "anatomically pointless," and "clinically irrelevant": "When on earth would your muscles ever do that naturally?" one male TA in the prosection course asked. "What is the point of showing that, except to show that you can do it?" a female dental student asked. The way Von Hagens and his team dissected and plastinated this cadaver puzzled most participants who mentioned it. There was, to their ways of knowing, no pedagogical relevance to the pose because it demonstrated nothing anatomically significant about the structure–function relationship nor would it offer the audience any useful understanding of their bodies.

FIGURE 5.1 "The Runner." Reprinted with permission from Gunther von Hagens's *Body Worlds* and the Institute for Plastination, www.bodyworlds.com.

Most interesting for my argument is the way participants deemed this body, unlike others in the exhibition, to be "ugly," "gross," and "weird." As we see in figure 5.1, this cadaver, though carefully and skillfully netterized, deviates from the classical and idealized aesthetic of the usual athletic pose. The dissection and arrangement of muscles in this deliberate posture suggests the visual conventions of an exploded diagram. In fact, the body itself appears to be an almost photographic glimpse of a human body exploding from the inside out. Ordered in this way, the cadaver is an unruly, abject model of a pose and a condition that defies the conventional structure–function association and resembles a body violently destroyed by dissection. The anatomist's choice to cut the muscles in certain locations and not others emphasizes the destructive work of dissection. Instead of exemplifying the unmaking of an unruly body into one made anatomically beautiful as the other plastinated cadavers seem to represent, "The Runner" seems to have been unmade even further by dissection.

The much-maligned plastinate, I argue, has an anatomical precedent, one that emphasizes the history of anatomical illustration and not contemporary clinical anatomy. "The Runner" is reminiscent of a more obscure but no less famous anatomical illustration from Paolo Mascagni's nineteenth-century European anatomical atlas *Anatomia Universale* (1833). (See figure 5.2.)

In the original plate created by anatomist Mascagni and artist Antonio Serantoni, the body is not posed as if running, but stands or lays with head turned and one hand raised as if gesturing for attention. Mascagni and Serantoni have rendered the major muscles of the cadaver in a disarticulated style similar to "The Runner," which was dissected, plastinated, and posed more than 100 years later. In this earlier display, some muscles of the forearm are even tangled around the cadaver's fingers, leading me to interpret this as a representation of a cadaver resting on a table in middissection. While "The Runner" may be (to participants of my study) "ugly" or "clinically irrelevant," it nonetheless recalls the intertwined historical traditions of anatomical dissection and image-making, of transforming the body through dissection into a presentational object that can then serve as the basis for anatomical illustrations. *Body Worlds* makes visible this mirroring relationship between anatomical illustration and dissection by exhibiting 3-D human models stylized as either famous anatomical illustrations or examples of structure–function relationships in action. What disturbs us about *Body Worlds*, if it disturbs us, is in part the way the exhibit allows the public to witness and to some extent participate in the anatomical vision of the gross anatomy lab.

From unruly chaos of the (supposed) natural body, the cadaver is unmade into a gruesome explosion of the (supposedly) unnatural body, one that exists only as a result of human intervention. Through its seeming violation of the anatomical-aesthetic sensibility this cadaver underscores the significance of *Body Worlds* and anatomical dissection, implications often troubling to a general audience and often unquestioned by an audience trained in anatomical vision. That is, the ordering practices of dissection unmake a physical body in order to make

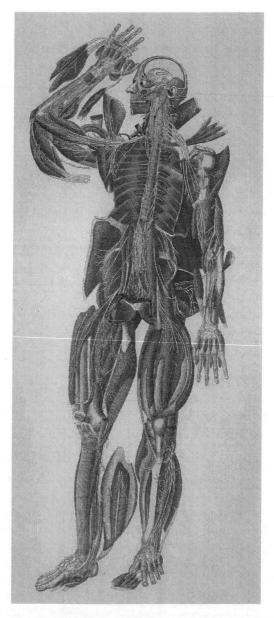

FIGURE 5.2 Plate from Paolo Mascagni's *Anatomia Universale* (1833). Reprinted with permission from the National Library of Medicine.

a conceptual body. Through the use of naturalistic images as mediating artifacts, dissection reveals and appreciates an aestheticized, netterized body that the undissected body can never attain. This process of making through unmaking, though difficult even for the dissector, is pedagogically, medically, and aesthetically significant.

Conclusion

Anatomical dissection is an embodied practice of looking, touching, and cutting that transforms abject, unruly dead bodies into aestheticized objects that instantiate and materialize the abstract anatomical body. Through the use of verbal texts and multimodal displays, primarily anatomical atlases, participants meticulously sculpt with forceps, scissors, and scalpel cadaveric flesh in order to reveal a particular set of structures. Through a dual process of revealing to learn and learning to reveal, students, TAs, and instructors engage in bodily, even intercorporeal, activities that require them to physically and manually maneuver through cadaveric bodies, excavating layer by layer to find, identify, and expose the anatomical structures in question. From atlases to whiteboards to the cadavers at the next table, mediating artifacts guide the distributed cognitive labor of dissection and, through these mediating artifacts, participants enact a 3-D display of anatomy from unruly cadaveric specimen. As enactive representations, atlases, and *Netter* in particular, provide the road map by which participants navigate, offering the way through the body's terrain as well as characteristic images of the final destination. By modeling their dissections after atlas plates, participants strive to make cadavers resemble these naturalistic displays of idealized, netterized bodies. Inevitably, dissection, with its intimate relationship to these atlases, entails ordering and making normative the abject bodies that participants come to understand as simultaneously aesthetic and scientific. The more carefully and skillfully participants can make the cadaver mirror *Netter*, the more useful and beautiful that cadaver becomes—beautiful as an emblematic example of the anatomical body. Yet, the cadaver's resistance to this poiesis complicates dissection as an ordering practice and an embodied skill. From anomalies to diseases to the difficulties engendered by a stiff dead body, the cadaver's recalcitrance unavoidably complicates this image-making work and encourages participants to appreciate the body's intricacy and the expert skill required to make it beautiful.

Netter and other multimodal displays, then, are tools and coding schemes for enacting anatomical presentations. Specifically, in dissecting, participants enact perceptual experiences and aesthetic frameworks made possible by their sensorimotor engagements with mediating artifacts. By engaging in the skilled and distributed activities of dissection, participants learn to make the anatomical body according to the stylized visions these atlases provide. Also, through their hands-on work at the microlevel of practice, participants develop a skilled perspective that is practical and conceptual—ontological, epistemological, and aesthetic. In the gross lab, students and TAs learn to recognize and appreciate the anatomical body by learning to make that body. Anatomical dissection not only trains their eyes, hands, and bodies to respond to cadavers in a manner structured by their budding expertise, but it also trains them to conceive of the body as a contradictory object and subject that is anatomical and philosophical, scientific and aesthetic, biological and personal.

In TPC, we engage in similar systems of distributed cognition that encourage in us ontological, epistemological, and aesthetic frameworks. Spinuzzi (2003,

2008), Swarts (2008), Van Ittersum (2009), and Winsor (2003) demonstrate that much of the labor involved in technical communication is distributed across an array of bodies, objects, and even locations. This distribution holds whether we are working in telecommunications, medicine, engineering, or simply typing on the screen in front of us. Doing this work, we engage with mediating artifacts, such as displays, documents, discourses, and objects, through which we inevitably enact sometimes subtle aesthetic categories that seem to us unexceptional. Take for example the much maligned yet actively deployed notion of clarity, also discussed as transparency. Researchers have criticized this concept as requiring a positivist approach to communication (Miller 1979), one based on a naïve conduit model (Slack, Miller, and Doak 2003) that inevitably conceals the complex ways technical communication shapes knowledge through language (Longo 2000). Yet our textbooks and syllabi continue to advocate some rendition of clarity, whether expressed in written prose, document design, or visual display.

For better or worse, our dogged insistence on clarity is not only a practical and rhetorical goal (based on audience and user concerns), but it is also an aesthetic category we have developed through our habituation with mediating artifacts that supposedly enact this clarity. From the elegance of Tufte's graphical displays (1997, 2001, 2006) to Ikea's equally loved and hated instructional documents, TPC has established its own aesthetic sensibility from collective, distributed, and embodied practices. Like the gross anatomy students, we too need to be aware of the potential consequences of this trained vision, a vision that involves technical, scientific, and medical displays and our embodied interactions with them. As I explore in the next two chapters, our trained vision is made possible by the rhetorical frameworks through which we learn to make sense of those images, objects, and practices. Both the way we talk about the objects of our work and the work itself shape our perceptions and beliefs about who we are and what we do.

6

DOWNPLAYING PERSONHOOD

Anatomical Focus and the Praise of Cadavers

Ask any gross anatomy student what role the human cadaver plays in learning, and you are likely to get a variety of answers that imply differing articulations of the body. Take for example, the words of the dental student, Randy:

> Sometimes you are just totally into what you're doing, and you kind of just view [the body] as a science experiment. And other times you step back and you realize, "wow," just what a phenomenal thing it is. It is just like the greatest machine, if you really look at how things are put together.

While Randy marvels at the body's intricacy, he acknowledges that this awareness is perspectival; you only come to it if you "step back" and "really look at how things are put together." For medical student Javier, the cadaver before him is more personal, "someone's grandmother, not just some body." Through his repeated use of the pronoun "her," Javier is careful to personify the cadaver as a person or former person. On the other hand, another medical student, Arthur, seems overwhelmed by the chaos of the cadaveric body and the difficult work of dissection. For him, the body is "like meat you were trying to make sense of." Later, perhaps out of a wish not to seem disrespectful, he corrects his perhaps coarse characterization and likens dissection to "chiseling a shape out of hard rock." A biological machine, a former person, or an object to be sculpted with surgical precision: in the gross anatomy lab, the cadaveric body is all these and more.

During our interview, the director of the dissection course, Kyle, lauded cadavers for their "realism, variation, and the learning that [comes] with actually dissecting." First, the cadaveric body, in his words, "injects a certain amount of anatomical realism into the education." No matter how detailed or precise, a naturalistic image, such as those in *Netter*, cannot offer the "three-dimensionality of being in the room present with the body." Second, the real bodies on display exhibit authentic anatomical variation that shows students how the body differs "from person to person." Lastly, as I discussed in previous chapters, students who engage in embodied observation and dissection learn the anatomical body by making that body visible in cadaveric flesh. The director's assessment of the importance of cadavers echoes similar sentiments expressed by every instructor and TA I interviewed officially during my fieldwork, as well as those I have spoken to casually since I left the labs. Most published accounts by anatomists agree (Drake et al. 2009; Gogalniceanu et al. 2008; Hanna and Freeston 2002). Ellis (2001), for example, extols the way dissection teaches students medical language, manual dexterity, the ability to communicate and use sources to confirm anatomical evidence, and "the wonders of biological variation" (149). He also claims that dissection can "acclimatize" students to "the reality of death" (150).

As empirical studies suggest, most anatomy students seem to feel the same; that is, working closely with cadavers either through dissections or observations of prosected bodies is beneficial to learning and generally a positive experience (Böckers et al. 2010; Dinsmore, Daugherty, and Zeitz 2001; McGarvey et al. 2001). Yet, as most participants in my study admitted, students can and do learn a great deal of anatomy from pictures and books. What these displays cannot offer, as I have discussed, is the complex three-dimensionality of the anatomical body, which participants in cadaver labs learn through direct exposure to human bodies. Gillingwater (2008) asserts that firsthand experiences with human material are necessary for deep (rather than superficial) anatomical learning. But do physicians, dentists, or nurses need this deep knowledge of the anatomical body?

This fundamental question confronted me early in my fieldwork. One afternoon I had lunch with a bioethicist who seemed deeply skeptical of the need for cadaveric anatomy, or at least dissection-based courses. From his own experience as a medical student, he found the gross lab an unnecessary and demoralizing experience, one that dehumanized students as much as the former living bodies among them. As sympathetically as possible, I explained how my experience of the labs, gained (at that point) from my pilot study and two semesters of unofficial observation, differed from his. The instructors I observed did not seek to humiliate students, nor did they encourage students to dehumanize cadavers. While neither of us persuaded the other, our conversation left me with a number of questions: Why were my experiences and those of many students and TAs so different from his? Why did so many participants view cadaveric anatomy as useful and often positive? How much of this belief stemmed from evidence gained through embodied practice, and how much of it was induced in us through overtly rhetorical means?

As an educational space and a domain of situated practice, the gross lab is structured through rhetorical discourses aimed at persuading participants and inducing in them certain ideas about anatomical education. After all, rhetoric as an induction to action and belief is, according to Burke ([1950] 1969), an "ingredient" in all forms of socialization (39). Rhetoric, whether expressed through discourse, documents, displays, objects, human bodies, and even institutional procedures, works by encouraging certain actions and perceptions in social agents and, in the process, it builds social bonds among agents who are moved by a particular message or worldview. To address these pressing questions, in the next two chapters I take up the role rhetorical discourses play in the development of trained vision and technical expertise.

So far, I have demonstrated how anatomy education in the labs I studied is made possible by the cadaveric body. Whether students actively dissect for themselves or view and touch what another has dissected, the dead body makes the anatomical one possible or at least visible. Time and again, participants describe these bodies of the once living as the real anatomy to which all other multimodal displays refer. Cadavers are the specimen material made anatomically beautiful by the work of dissection; they are the corporeal objects that train students in the embodied and rhetorical practices that are needed to recognize descriptive and relational evidence. These dead—along with the displays, tools, and practices used to make sense of them—help educate students in a body of medical knowledge while guiding them to see the body as medical knowledge in the flesh. I have argued that cadaveric bodies are not the only ones transformed by the practices of the gross lab; through their experiences, living participants are transformed as well. Now, I turn more directly to those processes of transformation. These processes are constituted not only through participants' embodied engagements with apodeictic displays, but also through the epideictic discourses and procedures that praise the cadaveric body and eulogize the anatomical donor as a gift giver. These epideictic formations of praise and memorial acculturate participants to the values and orthodoxy of anatomical science.

In this chapter, I examine how students in the anatomy laboratory come to see the body in a new way, one that merges the physical body we *are* with the anatomical body we *become*. I do so by exploring and amending the classic notion of clinical detachment, or the distancing mechanism that allows participants of the gross lab to allay the anxiety-producing nature of cadaveric anatomy. Through an analysis of my data, I argue that participants learn to navigate and even sustain the tension between the body as science and the body as person by adopting the powerful institutional rhetoric that praises as ideal cadaveric anatomy and dissection. This praise, or encomium, of cadavers structures participants' experience of working closely with cadavers by encouraging students and TAs to hyperfocus on the actions of dissection (as opposed to the object of those actions) and the scientific usefulness of the body. By influencing how participants make sense of cadavers, this encomiastic attachment to dissection sensitizes participants by making them more aware of and attentive to the body's biological usefulness. This sensitizing technique, what I call anatomical focus, is a rhetorical process of attunement by which participants learn to tune into the cadaver's biology and anatomical utility by focusing on the actions involved in dissection (viewing, touching, cutting) and not the former living person. By exploring how rhetoric structures the participants' relations to the cadavers and facilitates a certain formation of clinical detachment, this chapter also makes a more theoretical argument about the mutual construction of discourse and practice in the formation of technical expertise. That is, I want to suggest that rhetorical constructs operate through physical, even procedural means. Whether expressed through discourse, objects, or procedures, rhetoric structures our perceptual enactments, thus making possible certain ways of seeing, knowing, and being. Discourses, like objects, play a significant role in enactive cognition.

Epideixis, Cultural Values, and Anatomical Education

Epideixis, whether exhibited in a speech, a text, or a set of procedures, creates and maintains a culture's system of values (Sullivan 1991, 229), and it does so by determining what is and is not appropriate and possible for that culture. Thus, epideictic rhetoric is more than merely a ceremonial discourse of praise and blame, as Aristotle's *Rhetoric* (2007) so powerfully and lastingly categorized it. In his criticism of this Aristotelian reduction, Rosenfield (1980) characterizes epideictic rhetoric's primary function as "acknowledgment" and "disparagement," specifically the "recognition" or the "refusal to recognize" certain ideals as meaningful and worthy of affirmation (133). For ancient Greeks, particularly Athenians, epideictic demonstrations provided serious opportunities for social formation, political agency, and cultural indoctrination (Carey 2007, 237). By praising or acknowledging certain actions and not others, the speaker implicitly advises the audience to undertake or uphold those actions. After all, praising and advising are closely related, a fact Quintilian, in *Institutio oratoria*, understood when he connected exhortation with exaltation (117).

According to Habinek (2005), the goal of a speech such as a panegyric or encomium was "not to capture or reflect a preexistent reality," but instead to compose or perform "a kind of truth," specifically, "an unforgettable and socially significant vision" of reality (54). These performances, which might include speeches and poetry, seemed to operate by way of a constitutive rhetoric that sought to create as a "socially valid and meaningful entity" the institutionalization of a certain vision of the world (54). A funeral oration (epitaphios logos) eulogizing the selflessness of a fallen soldier not only articulated certain actions as virtuous and others as shameful, but also supported cultural, institutional, and individual practices that perpetuate this ideal. It did so by creating "feelings of emulation" that lead to imitation and action (Sullivan 1993b, 118). The audience of epideixis, then, is invited not to judge but bear witness, "to be a *theoros*" to this act of making (Rosenfield 1980, 140, italics in original). This form of theorizing or understanding, which Oravec (1976) deems a type of observation, ultimately prepares the audience for learning and action (166). Epideixis, then as now, is "essential for acclimating" participants to actions deemed appropriate to the orthodoxy of their community (Hauser 1999, 17). Like ritual, epideixis creates and fosters community, in part by offering guidance in how we should conduct our lives (Carter 1991, 213). All forms of education, even scientific training, operate by way of an epideictic impulse that teaches participants the ways of thinking, seeing, and being deemed appropriate to that discipline (Sullivan 1993a). This trained vision involves inculcating students into certain modes of reason and feeling, and these modes allow them to internalize what Sullivan (1993a) describes as "the ethos of the culture" (71).

In the history of anatomy, no epideictic formation has been more powerful than the exaltation of the cadaveric body as the ultimate anatomical object. As far back as second-century Rome, Galen argued that anatomical knowledge should be the basis of medicine (Cunningham 1997, 26). Unfortunately, because of legal prohibitions, Galen's own knowledge of anatomy was based largely on his dissections of rhesus monkeys, sheep, pigs, and goats (Nutton 2002, 800–801). While Galen did get firsthand glimpses of human anatomy during his time caring for gladiators in Pergamum (157 CE) and as a practicing physician, his at times cursory anatomical observations came primarily from looking at skeletons, the surface anatomy of slaves, the interiors of executed criminals, and the bodies of wartime causalities (Nutton 2004, 231). Still, Galen espoused the benefits of anatomical knowledge in such second-century works as *De usu partium corporis humani* (in print since 1528) and *De anatomicis administrationibus* (in print since 1531), even while his own knowledge was flawed by incorrect comparisons of humans to animals and an overreliance on Platonic conceptions of the body (Cunningham 1997, 29; Nutton 2004, 230–31).

Galen's ideas continued to hold sway over anatomy education in Europe until the fifteenth century when anatomical dissections became public performances (Cunningham 1997). During these early-modern public dissections, the professor (*lector*), seated atop an ornate chair (the *cathedra*), would read aloud from some authoritative text, while leaving the messy work of dissection to the *sector*, often a barber or hired dissector whose cuts were meant to correlate with the recited passages. In order to direct the students' attention and the sector's labor, an *ostensor*, or demonstrator, would stand close to the open body and indicate with a long pointer which structures were being discussed and which should be cut next (Carlino [1994] 1999, 12). These apodeictic performances, focused more on hearing the anatomist's recitations than on seeing the body, were slowly eroded by increasing calls, as in Mondino de' Liuzzi's *Anathomia* (completed in 1316 and published ca. 1474), for more empirical observations and hands-on pedagogy. Nearly two centuries after Liuzzi's work, anatomical texts by Benedetti (1502), Berengario (1521), and Massa (1536) praised cadaveric anatomy and corrected some of Galen's errors based on evidence the author-anatomists gleaned from human dissection (Carlino [1994] 1999; Cunningham 1997, 59). As Park (2006) points out, the final chapter of Benedetti's work is even titled "In Praise of Dissection"; in epideictic prose, Benedetti exhorts medical students and physicians to "frequent anatomical dissections" so they might "see the truth" in the "works of nature" (166–67).

However, it was Vesalius, working in the 1540s, whom historians often credit with stirring the stagnant waters of early-modern medical education. While a professor at the University of Bologna, Vesalius refused the honors of the cathedra and the aid of the ostensor and the sector in order to perform public dissections himself before large audiences. Vesalius's years of anatomical study and performance culminated in the publication of *De humani corporis fabrica* (1543), an exhaustive seven-volume work that furthered his corrections of Galenic errors and advocated the necessity of cadaveric dissection for understanding anatomy and training physicians. The publication of the *De humani corporis fabrica*, with its beautiful and realistic images of dissections and human structures, led to debates that changed anatomy education, irrevocably turning it toward the human body and embodied experience and away from oral descriptions and textual practices of reading (Carlino [1994] 1999; Cunningham 1997; French 1999).

In the contemporary gross lab, this epideictic formation that praises cadaveric study lives on and makes possible all laboratory activities. The encomiastic attachment to the cadaver is expressed explicitly through verbal utterances and more implicitly through the practices and procedures that structure the entire educational enterprise. This encomium permeates the laboratory, authorizing the practices involved in cadaveric dissection, demonstration, and observation. For instance, whenever a TA, an instructor, or a student praises the unique benefits of the cadaveric body or marvels at its supposed wonders, that participant is not only recognizing the uniqueness of the cadaver as an anatomical tool, but is also affirming the values of anatomy education that prize the cadaver as necessary. Though I mentioned them briefly in chapter 5, I want to quote in detail the words of Miles as he explains to me the purpose of dissection:

> I think [dissection] helps play a role also in realizing how much precision it takes to find certain things. It's really difficult to find particular spots in the body and also to appreciate how complex the human body is. You know, all the Honda engineers could work for a century and still would not be able to design everything in the body. (Miles, dental student)

Echoing the sentiments of his mentors and teachers, Miles describes dissection as a process that fosters appreciation of the body and knowledge of its biological complexity. For Miles, this appreciation is expressed in mechanistic metaphors of the body as a machine as well as metaphors of the marvelous that describe the body as an object of wonder. Through this training in complexity, participants come to view the cadaveric body and cadaveric dissection as indispensable to medical training. This appreciation of the body's intricacy encourages students to take the course and dissections seriously, supporting the TAs' and instructors' suggestions to study daily. After all, students will need to know anatomy not just to pass the course but also to pass the licensure exams and to practice medicine. Though most anatomists and medical students do find the cadaver a useful tool, this rhetorical formation that extols its usefulness plays a key role in helping students adjust to cadaveric dissections. In particular, the instructors deploy this encomiastic formation not only to introduce students to the cultural values of medicine but also to help the students deal with potentially troubling aspects of cadaveric anatomy and the uncertainty of medical education.

Cadaveric Bodies and Medical Uncertainty

Fox (1979) recognizes three types of uncertainty that medical students face during their training. The first stems from "incomplete or imperfect mastery of available knowledge" (21) that involves confusion over what students should know and how they can best learn it (22). The second results from the "limitations" of existing (and knowable) medical knowledge, namely that even the best educated and the most experienced physician will also possess an incomplete mastery of knowledge, not to mention that fact that knowledge about research and clinical practice is never finalized but involves ongoing inquiry (22–23). The third form of uncertainty derives from the challenge of distinguishing the two— differentiating one's own lack of knowledge from the current state of knowledge in the field (21). As a result, medical knowledge entails a "training for uncertainty" in which students are encouraged to recognize, acknowledge, and tolerate uncertainty as an unavoidable consequence of training and practice (25). The instructors and curriculum play a formative and not always positive role in this training. For example, instructors can contribute to students' feelings of ignorance and uncertainty when students consider their knowledge in relation to their professors (23). But these same instructors also use uncertainty as a productive tool by showing students how they incorporate it into "experimental work" and methods of inquiry (25). For instance, when tagging cadavers, instructors model for TAs the types of deliberation needed to verify the efficacy of a tag.

The cadavers are objects of uncertainty that present students and even TAs with anatomical structures they must identify and know, on bodies that complicate and at times resist these practices of identification. These bodies also offer one of the starkest contradictions in medical education. Students are required to closely interact with the dead in order to benefit the living. That is, the dead bodies they eviscerate through dissection offer knowledge the students will use some day to help save living bodies. This contradiction is by no means avoided in the gross labs I observed; instead, TAs and instructors repeatedly address the cadaver's conflicting status as a valued instantiation of anatomical knowledge and a former living person. After all, medical educators want their students and residents to possess, in the words of Good and Good (1993), "competence" and "caring."

Competence requires medical knowledge and know-how, while caring involves "attitudes, compassion, and empathy" (91). The first seems to require objectivity, and the second sensitivity. Walter (2004) views this dual impulse as "a latent function" of anatomical dissection, namely that students are trained to "suspend personal feeling" and "see, discuss, and treat" the human body with "objectivity" (464). Horsley (2010) offers a similar sentiment when she describes anatomical education as a kind of balancing act; participants find themselves "confronting and juggling" everyday reactions to a dissected dead body with their growing clinical perception of that body as an anatomical object, resource, and instrument (10).

Fox (2000) identifies three strategies students use to deal with the "multiple uncertainties of medicine," all of which play upon this balancing act of objectivity and sensitivity. The first is a process of "intellectualization" in which students seek a mastery of uncertainty through a greater command of medical knowledge and skill (410). The second is the development of "a more detached kind of concern" in which students mute the awareness of uncertainty and the more "emotionally evocative" aspects of medical education such as the presence of death and dying (411). This second strategy of "detached concern" (Lief and Fox 1963, 12) is, I contend, a particularly important rhetorical formation that structures how participants perceive of cadavers. Lastly, the third strategy is the adoption of a specific form of "medical humor," including both gallows humor and "self-mockery" (Fox 2000, 411). In the gross lab, participants implement each of these techniques in differing ways. They do so not just to mitigate uncertainty, but also to make peace with their close connection to cadavers. Intellectualization or the objectification of the body along with humor and moments of storytelling, I argue, are components of detached concern, which, in the gross lab at least, is made possible through epideictic discourses that exalt the cadaver's novelty and epideictic procedures that allow students to attend to its anatomical utility.

Detached Concern and the Anatomy Lab

Detached concern, as Fox (1979) defines it, is the process whereby medical students learn "to combine the counterattitudes of detachment and concern" in order to strike a balance between the objectivity and empathy required to practice medicine. Fox articulates this form of clinical detachment in relation to students' work with cadaveric bodies. In order to perform an autopsy, students are "called upon to meet death" with the "impersonal attitude of scientists." Yet students are also expected to remain sensitive to the "human implication" of their work (56). Detached concern resembles Horsley's (2010) explanation of clinical detachment as "the muting of one's expected responses" to the dead that allow anatomy students a way of emotionally and intellectually resolving the irony that they must cut into one body to later aid another (8).

In her famous history of anatomical practices in nineteenth-century England, Richardson (2000a) speaks of the anatomist's clinical detachment as "a defensive barrier" that allows him or her to perform tasks that would otherwise be "taboo or emotionally repugnant" (31). Clinical detachment juxtaposes "objectivity" and "emotionlessness" (Richardson 2000b, 31), with the assumption that medical students learn to view the body with clinical objectivity by suppressing or mitigating their emotional reaction.

Many participants, however, were troubled by the implications of this detached concern, particularly those who felt that emotional responses to cadaver should not be blocked but instead addressed. One medical student voiced his concern about the consequences of his feeling of detachment: "I wonder if we aren't losing part of our humanity by doing this." His words echo those of William Hunter, an eighteenth-century British surgeon who advocated a "necessary inhumanity" among those training to be doctors, particularly when it came to dissection (Richardson 2000b, 104). Richardson (2000b) explains that Hunter's words do not represent a callous lack of empathy but instead a need for training surgeons to be quick, skilled, and precise—capacities they could never develop if they were squeamish about dissection and cadavers. Because of its implications, Richardson even prefers Hunter's necessary inhumanity to clinical detachment: "The phrase has more precision, a suggestion of calibration, even a hint of warning, which clinical 'detachment' and 'objectivity' lack" (104). Gross anatomy students need only the emotional distance necessary to work on and with the cadaver, and no more.

Richardson's evocation of Hunter's words implies that clinical detachment in the gross lab need not preclude empathy or an empathetic connection to cadavers and the anatomical body. In a letter to the editor of *Anatomical Sciences Education*, Hildebrandt (2010) seems to agree. Responding to an empirical study by Böckers et al. (2010), Hildebrandt (2010) suggests "an alternative interpretation" of their findings that "dissection in the gross anatomy course is not suitable for encouraging an early development of empathy." Hildebrandt concedes that the study finds students entered the gross lab course "expecting to develop empathy" along with other professional skills, and that students at the end of the course reported they had not "done so to the degree they had expected." But in a move that questions the dichotomy of clinical detachment and empathy, Hildebrandt asks whether students' viewed the development of clinical detachment as a decrease in empathy and whether such a decrease actually occurred. This question leads her to speculate how the findings might have been different had students been made "aware of the factor of clinical detachment" as an inevitable result that need not preclude empathy (216).

In other words, can gross anatomy students come to understand clinical detachment and empathy as two necessary and nonconflicting states of being? Can the experience of the gross lab facilitate a kind of detachment that is neither emotionless nor depersonalizing? By introducing students to the issue of clinical detachment, can professors and administrators shape students' empathetic engagements? Based on my field notes, interviews, and my own experience in the labs, I find that students can and do exhibit a form of detached concern that includes a palpable element of empathy. This type of clinical detachment, what I call anatomical focus, involves a "necessary inhumanity" that operates through a twin impulse of distance and closeness: a distance from the emotional content of the cadaver as a dead body and a closeness to the anatomical usefulness of that same body. Anatomical focus is a sensitization to the cadaver as an object of anatomical study as well as a former person and an anatomical gift. While in the next chapter I discuss the empathetic connections fostered by the analogy of the gift, here I illustrate the ways that epideictic discourse that praises the utility and necessity of cadavers supports anatomical focus. I examine how this rhetorical formation inevitably objectifies the cadaveric body by downplaying the individual personhood of that body.

Objectifying the Cadaver: Downplaying Personhood

In the labs, clinical detachment entails a careful balance of detached objectivity and emotional sensitivity, juxtaposing but not eradicating the human body as an object and a subject—or as biology and personhood. Take this dental student's discussion of his perceptional shift from cadaver as person to cadaver as tool:

> The first day I thought it might be a little weird. Once we opened the bodies up, you know, raised them up and then looked at them, it was a little bit [weird]. But once I started cutting, it became all science to me, like nothing. I didn't have any barrier, any emotional barrier to cross at all. It was just 100% science and time to learn, so I did just totally fine. (Carl, dental student)

Carl recalls that his trepidation and anxiety arises, as for most participants, during his anticipation of the first encounter. For Carl, this trepidation soon gives way to a kind of removal of personhood—the body becomes a scientific object, "100% science." The emotional barrier recedes as the emotional content of the cadaver recedes, or is reduced, through this process of rendering the physical body as the anatomical body.

One course instructor provides a detailed account of the movement toward what Fox (1979) terms a "scientific orientation" towards the body (65):

> You're so wrapped up in identifying muscles that it's no longer—I mean, you still appreciate that it is human. I mean, you still have the respect. But your mindset is, you know, "I have to learn this muscle. I have to learn this intestine." And so the—I think it is a concentration on the learning to overcome your inhibitions about a dead body.

He continues,

> But I mean, it goes from the students are afraid to get close enough to even touch, and they are hesitant to look. But two or three weeks later, you know,

they are up to their elbows moving organs around. So it's an odd phenomenon, but it's just, ah, it's kind of a switch that goes off, you know. Before you know it, you are just fully stuck into learning. (William, instructor of both courses)

First, he describes how participants' specific focus on the course objectives and lab practices encourages them to home in on specific structures and not the overall bodies. However, as he argues and as I discuss later, this focus on structure does not preclude a respect for the personhood of cadavers, particularly a respect for the gift of cadaveric donation that made this educational experience possible. Second, once students make this transition, they become, in his words, "fully stuck into learning." This kinesthetic and embodied metaphor describes the way students push themselves bodily into the content of the course (anatomical knowledge) and the learning tools of the course (human cadavers). For participants, anatomical focus is more than merely forgetting the personhood of the cadaver or even removing themselves emotionally from the task (what Carl above termed an "emotional barrier"). Instead, anatomical focus involves a perceptional shift that allows them to (1) simultaneously view the body as person and specimen, as well as (2) carefully engage with the physicality of the cadaver while disengaging from the emotional content of working on the dead. This perceptual shift, or rhetorical frame, is made possible by epideictic discourse that praises cadaveric dissection as fundamental to anatomical learning and the medical school experience.

On the surface, students deal with dissection by distancing themselves from the body's personhood; they do this by focusing on the anatomical body materialized in the cadaver. But clinical detachment in the gross lab is much more nuanced. Participants distance themselves from the cadaveric body's reference to its former personhood through a focus on the process and not the object of that process. This focus on process, a kind of process orientation, manifests first as a general attention to their larger career trajectory and second as a hyperfocus on the manual tasks operated on specimen tissue or the body in parts. This focused distance, as I call it, inevitably emphasizes the cadaver's biological value as an authentic materialization of the anatomical body. Here the once-living body becomes "all science"; yet it is the cadaver's role as real anatomy, as the ideal anatomical body, that subtly acknowledges the body as a corpse, as a dead person.

In other words, rather than displace the former personhood of the body, this articulation of clinical detachment often experienced in the gross lab reemphasizes the anatomical novelty and utility of the cadaver. This manifestation of clinical distance invites participants to focus on activities involved in cadaveric demonstration, observation, and dissection. This process orientation offers the emotional barrier seemingly provided by the ethos of objectivity that provisionally deemphasizes the cadaver's individual personhood. As a rhetorical frame, this process orientation does involve a degree of distancing and depersonalizing.

But it does so in part by contextualizing the cadaver according to the encomium of dissection. That is, this process orientation allows participants to downplay the cadaver's former personhood by amplifying the cadaver's extraordinary anatomical value. This process focus and this heightened recognition of anatomical values are only possible because of the epideictic discourses and procedures that structure anatomy education. The traces of these rhetorical and embodied processes are exhibited in the ways students (1) engage the darker moments, (2) experience the focused distance necessary for dissection and observation, and (3) attend to the cadaver's unique biological value.

Engaging the Darker Moments

Many of my interview participants spoke of their initial apprehension of the first lab. Students in both courses were concerned about what they might encounter in a cadaver lab and about how they might react to that encounter. Would they feel lightheaded or nauseated? Could they handle it? Some also mentioned concerns their close friends and partners expressed about how the anatomy lab experience might change them. These apprehensions are understandable given our dominant cultural narratives of the anatomy lab, particularly popular accounts that describe gross labs as just that: gross. In their analysis of representations of human anatomy teaching and research in the news media, Regan de Bere and Petersen (2006), find that most coverage is framed predominantly as stories of "awe and amusement" or "fear and revulsion" (81). Anatomy labs either represent "miracles of modern science" that offer "gifts of life," or a "Frankenstein" world of "social and genetic engineering," in which anatomical dissection and research are depicted more as "rapes of the body" (82–83). There are disturbing moments in the long history of anatomy, such as the 1828 trial of Burke and Hare, who killed to provide bodies for the Edinburgh anatomy professor Robert Knox (Richardson 2000b, 133–40), or Pernkopf's *Topographische Anatomie des Menschen* (1937), with illustrations likely modeled on the bodies of victims of the Nazi Holocaust (Hildebrandt 2006, 92). Though it would be irresponsible to deny these crimes, it would be unethical to assume that the field of anatomy is not coming to grips with this history in ways large and small (Hildebrandt 2006, 2008, 2009a, 2009b, 2009c). Again, Hafferty and Franks (1994) identify a greater awareness of issues of ethics, death, and mortality as part of "the hidden curriculum" of medical education, made possible in part by students' close proximity to the dead (861).

One way to help students deal with these darker moments is to encourage students to acknowledge them in ways deemed meaningful by instructors, TAs, and other students. Fostering a sense of shared experience, often through storytelling, openly engages participants in the sadder or more troubling aspects in ways that support the ongoing epideictic formations that slowly socialize students into the cultural values of anatomy education. On the first day of the dissection course, additional third-year medical students were present to physically assist in dissection and offer support, advice, and understanding during students' initial cadaveric encounters. These epideictic opportunities to share and build a mutual trust are also moments of socialization and habituation, in that students come to better understand, as one female medical student put it, "how the TAs did it, handled it, the cadavers I mean, on day one." Most participants described these additional advanced helpers as good for the students and the regular TAs, easing students' anxiety and fostering a sort of "camaraderie" (a word used by several TAs and students). These epideictic moments introduce new participants to the social rituals of the gross lab, not the least of which is a better understanding of dissection and appreciation for cadavers. Through storytelling, TAs and instructors reflect on and share viewpoints and values they have developed through their interactions with cadavers.

These moments of storytelling are common in the lab. Lella and Pawluch (1988) and Reifler (1996) explore students' use of narrative to make sense of and make peace with the more gruesome experiences of dissection. Lella and Pawluch and Reifler even advocate the use of writing assignments such as personal narratives to help students cope. Narrative, as the vast research in this area illustrates, helps structure our experiences of illness and health (Segal 2005). Case narratives also play a significant role in the development of a practitioner's clinical judgment (Montgomery 2006) as well as the historical progression of medicine as a field of practice (Berkenkotter 2008). The movement toward narrative medicine, popularized by Charon (2006), encourages health practitioners to take seriously patient narratives because they can enrich clinical diagnosis and care. In the gross lab, storytelling is an epideictic practice that encourages a sense of community among participants, while at the same time transmitting the cultural values of anatomical education. Specifically, as was exhibited on the first day of the dissection course, these stories are used to acknowledge emotional and physical difficulties and to introduce students to acceptable responses to those experiences. But instructors and TAs have to acknowledge the darker moments before they can help students deal with them.

Hafferty (1991) uses the term "cadaver stories" to describe the "usually apocryphal" tales of medical students' comic pranks using cadavers or their parts (55). These tales serve a number of functions. For example, they introduce students to the lab's experiences and norms and guard against an excessively "emotional reaction to the lab" that might make a student seem "unfit to become a physician" (55–62). During my fieldwork, I witnessed no pranks nor heard any jokes involving cadavers, primarily because the instructors and TAs expressly and repeatedly discouraged such behavior. On the first day of both courses, instructors warned students not to "disrespect" either the bodies "or each other," in the words of one dissection course instructor. The absence of pranks, however, does not mean an absence of cadaver stories. TAs and students often shared brief tales of their own or others' reaction to iconic moments of cadaveric anatomy. One medical student, Donnie, endured some light and good-natured teasing from his team because he did not like coming to their evenings study sessions in the lab. "What?" he said, smiling, "It's a little creepy." Jack, one of the prosection course TAs who often came to the labs alone to do last-minute setup, described "having to get comfortable being alone in there." Elizabeth, a TA for the dissection course, confessed that while she was a student in the gross course, she once dreamed she ate a piece of her cadaver during a study session. Grimacing as she recounted the story, Elizabeth explained that it was probably a dream that literalized the idea of ingesting and digesting anatomical knowledge. Even so, she said, she became a vegetarian for the rest of the term. In each of these examples, participants light-heartedly admit to common gross anatomy experiences. The telling and retelling of these tales strengthens social bonds among participants who empathize with such experiences and reactions.

During my fieldwork, I did witness many participants' attempt to seem objective and unmoved by the lab experience. At least twice a week during the first three weeks of the dissection course, I noticed students stepping out into the hallway for air or a slow walk down the corridor. While no doubt students were taking a break from the physically exacting work of dissection, a few did seem to be stepping back from the emotionally taxing work of touching, cutting, and excavating bodies. On the afternoon that one team discovered cancer in the body of their female cadaver, I saw quite a few students taking breaks and getting air. Feeling overwhelmed and saddened, I too stepped out into the hallway, where I passed a student heading back into the lab. "Are you O.K.?" he asked me. "Yes," I replied and thanked him. "It was the lady with the cancer, I think," I added. "Yeah," he agreed. "She really got to a lot of us today." Even though I was an outsider to the community formed by instructors, students, and TAs, this student took a moment to check in with me because of our shared experience of the gross lab. Many of my interview participants praised the "maturity" (one student's word) of their peers who seemed to not only respect the cadavers, but also each other's ways of dealing with the darker or sadder moments. Body-buddy teams usually checked in with each other when a member seemed particularly quiet or distracted. In the words of one student, many of them had "honest talks" with their peers about "how we feel about doing this."

Experiencing Focused Distance

Eventually, students grow accustomed to working closely with cadavers. Participants in both courses expressed surprise at their ability to adapt quickly to interacting with these formerly living people. Miranda's description of this process is emblematic and illuminating:

> But I am surprised at how quickly you adapt. And, you know, it is like, you know, you just have to do this, and then eventually it just becomes—You

try and put that distance there so that it becomes about finding the structure as opposed to, you know, you're cutting into a person. But it happened so fast, you know. About five minutes into it, I think, we all kind of felt, "O.K. We're going now, and we are looking for muscles." (Miranda, medical student)

This "distance," as Miranda refers to it, is "put" there by the students themselves in order to cut into another human body. Related to the affordance of the cadaver as viewable, dissecting the body becomes a way of finding structures. She describes a kind of self-navigation through the body that operates as a distancing technique, as students look into another body for muscles that allowed that body to function. Participants were surprised not by the development of this coping mechanism, but instead by how quickly they developed it (or how quickly it developed in them).

Many students explained this distance as a tendency to separate themselves from the personhood of the body by focusing on the larger activities of the course. Some of them described it as a kind of second-nature adaptability:

But then as the course progressed, it became like second nature. Because I didn't think about it. I would go—When I would touch everything, dig around in everything, I wouldn't think about it. And then right after I would go do something else, and I wouldn't think about that. And I would think that that was the human body, that this was the cadaver, you know. That I'm actually digging around in here and pushing in this. (Rodney, undergraduate student)

Rodney's use of the term "second nature" is interesting in that it acknowledges a developing dispositional tendency, in this case a type of habituated vision. Rodney and other students become so immersed in the task through the repeated exposure to the body as tool that the embodied practices of anatomical vision become second nature. Becoming skilled at a task, according to Dreyfus (2005), involves no longer needing to consciously remember or consider performance-based rules and steps. In the lab, this second-nature adaptation decreases the emotional difficulty or apprehension. However, this habituation is not without moments of tension, especially when the humanity of the cadaver interrupts this distancing operation.

Horsley (2010) understands this distancing mechanism as part of the "immersion factors" (11) of working in the gross lab, specifically the "practical and cognitive demands" that inevitably narrow participants' focus so they can "achieve the necessary distance" to perform their work (12). This emotional and intellectual distance from cadavers as dead bodies encourages a renewed focus on the tasks being performed and not the object of those tasks. This focused distance operates on two levels. First, by focusing on their larger career objectives, as future doctors, participants understand dissection as a task crucial to their intellectual and emotional development of a particular ethos:

> It was just realizing that I need to be comfortable with this now because I'm going to be dealing with this for the rest of my life, because I wanted to go into medicine, and I want to go into surgery. And so if I cannot handle this now, how can I handle this in the future? (Steven, undergraduate student)

Steven expresses a sentiment common to many students and TAs, especially TAs in the prosection course. Getting used to cadavers is an occupational requirement. Connected to this career focus is the way many medical and dental students understand working closely with their peers as mitigating moments of unpleasantness. One medical student, Daniel, explains how their common career goals and the camaraderie of the experience can offer emotional support:

> I think the best way that we get used to it is kind of like the military experience, that you are going through it with other people, so you can discuss this with other people as opposed to—. I mean, you can relate to other people as opposed to if you were just doing this yourself. You have a lot of things that, you know, you would really want to talk about with people, but you couldn't relate it to anybody else. So having body bodies and an entire 165-student class to relate to makes it a lot easier.

As Daniel's comments reveal, this camaraderie and the ability to talk with peers help acclimatize students to the course and the cadavers. Daniel's comparison of medical school to the military stems, no doubt, from his own background in the military; even still, the gross anatomy experience serves a similar initiation ritual for medicine and dentistry that boot camp serves for the military. Both are intense experiences perceived as life-changing rites of passage that introduce initiates to the practices, values, and norms of that culture. Simultaneously, these experiences immerse students in embodied rhetorical actions that allow them to make sense of, make meaning from, and engage with the assemblage of displays, objects, discourses, and practices of that culture.

Second, this distancing mechanism is evident in the way participants come to focus on the specific techniques and gestures involved in dissecting:

> But then again, we started working, and they became more about finding the anatomy than it was about the person, herself. So, you know, it's definitely keeping the purpose of what you're doing in mind rather than focusing on what you're doing to someone that used to be living. I mean, so, yeah, it's definitely more of an object to work with than a human for me. (Samuel, medical student)

For Samuel, this focus on doing, a preoccupation with practice, allows him to attend to the body as an object of that doing. The body becomes an object subsumed by the practices acted upon it—namely, a collection of structures one must find, touch, and often dissect—with the emphasis placed on the ordering processes (dissecting, identifying, and demonstrating) that give the "object" (the body) its meaning and coherence. This focus on practice does not necessarily mean that students lose sight of the object's humanity. As another medical student, Louis, describes it, "I didn't forget that it was a real person, but it [dissection] became more of just something that I'm doing, rather than something I'm doing to someone."

There is a kind of "but-still" aspect to this focused distance, with which participants home in on the particular structures of the cadaver, searching for the structure in question, without completely losing sight of the social and personal nature of the cadaver as a dead person. It is this but-still nature of cadavers that encourages a hyperfocus on task as students perceptually train their lens on the body in parts and do so in a way that incorporates an aesthetic appreciation and curiosity. Hirschauer (1991) finds a similar focused distance in the way a surgeon's view of the body is narrowed to the site of incision by the draping of cloth that inevitably distances the surgeon from the personhood of the patient as well as the rest of the patient's body (289–90).

Here, another student discusses this tendency to hyperfocus on actions and not objects:

> You kind of just have to focus on the idea that "I need to learn this and this and this." And this is the best way for me to learn this, by doing this, and that's the way you look at it after a while. Rather than "I'm cutting off somebody's this." But instead, "O.K. For me to see this, I have to do this. And this makes sense." (Gordon, dental student)

This hyperfocus on doing can become a long-term adjustment strategy perhaps unavoidably encouraged by the lab's embodied practices. After all, students have to walk from table to table, finding and identifying structures in bodies that do not always resemble the orderly images. To orient themselves to the cadaver, students focus on some prominent feature, as one undergraduate comically illustrated one day when she asked a female TA for help because "I can't even figure out which end is which." Though this is an exaggerated and atypical confession on the part of a student, the TA's reply is telling: "Focus on finding a landmark that you know completely." This TA suggests locating an obvious orienting landmark such as the head or the genitalia so that the student can make sense of the body and find the structures in question. This hyperfocus becomes part of how students work with cadavers because it makes the work emotionally easier; it contributes to the process of anatomical focus, and it mimics the type of activities they must perform during practical exams.

But for some students hyperfocus brings with it fears of losing touch with the cadaver's humanity:

> I walk over to a body, and I don't even think about the fact that I'm looking at their leg. I'm just looking for arteries, you know. I think you forget what you're actually—. I mean, you forget that this was actually a human at one time. (Amelia, dental student)

Amelia even compared the way students grow accustomed to cadavers to the way residents of urban areas learn to tune out the noise of sirens and car alarms: "You suddenly stop hearing it." TAs in both classes mentioned developing perhaps "too much comfort with" the cadavers, retelling stories in which students initially "looked pretty squeamish" whenever a TA would rest her hand or arm on a cadaver or even touch a cadaver in a casual way without pointing at structures. For TAs, this close and repeated proximity to these bodies seems to bring the same embodied and kinesthetic tendencies one experiences with other nonhuman objects. In other words, they touch the cadaver as they would any object with which they have grown comfortable.

Focused distance, which sharpens their attention to actions and reduces the cadaver to an object, does not completely eradicate the cadaver's humanity, only its individuality. When I asked Amelia to explain her siren comparison, she said that you stop seeing the "single person on the table," and you start seeing "the person as a type of thing, you know, the possible person and not one individual." This way of seeing the individual body as an example of a type illustrates, for Daston and Galison (2007), the development of trained judgment (311). Specifically, they argue that this form of objectivity trains budding experts to see and interpret visual displays as an example of a larger classificatory family (371). This focused distance, which allows students to foreground the processes they perform on the cadavers and background the former personhood of that object, is a component of their burgeoning expertise. This hyperfocus is supported by and even dependent on the lab's epideictic discourses that praise the unique experience of cadaveric anatomy. By taking up this encomiastic perspective, students and TAs are given a kind of necessary permission to zero in on the body's anatomical extraordinariness in order to perform firsthand observations and dissections on these once-living bodies.

Focused distance, I argue, operates as a type of epideictic procedure, a physical process that exerts a persuasive and ontological force that socializes participants to believe and, in this case, respond according to a markedly epideictic formation of recognition and acknowledgment, praise and valuation. I conceive of this procedural epideixis as involving what Burke ([1950] 1969) describes as "administrative rhetoric," or the use of physical actions (or "nonverbal" and "nonsymbolic conditions") as a persuasive display or appeal (161). Yet it also involves what

Bogost (2007) terms "procedural rhetoric," namely the performance of procedures and operations that persuade individuals toward certain actions. While Bogost's original term is tied to the operations that computers seem to invite of users, his concept of the way we are persuaded to perform certain tasks based on the affordances of the objects involved speaks powerfully to the suasory force of social actions. In the gross lab, these procedures function as epideixis in that they are supported by epideictic discourses and formations and induce in participants an uptake of those forms of epideixis.

Attending to Cadaveric Biovalue

Through epideictic demonstrations and apodeictic displays, participants of the gross lab develop a particular perspective that is (like all perception) embodied and rhetorical. That is, students, TAs, and instructors develop this way of seeing, thinking, moving, and being as a result of their immersion in the lab's embodied and rhetorical practices. This overall trained vision—anatomical vision—fosters an articulation of the dead body as anatomical specimen, an instrumental object of demonstration and verification composed of human flesh. This rendering practice is an unavoidable consequence of interacting with cadavers in this particular space because of the purpose of these bodies: to learn and teach anatomical knowledge by way of the authentic anatomy provided by the human body. The encomiastic discourses and procedures support the embodied practices of demonstration, observation, and dissection by giving those practices a larger cultural meaning.

As a result of this entanglement of rhetorical and embodied practices, participants of the gross lab come to see cadavers as infused with what Waldby (2000) terms "biovalue," or "a surplus value of vitality and instrumental knowledge" that is "placed at the disposal of the human subject." For Waldby, biovalue is produced whenever human material, or "marginal forms of vitality" (such as "foetal [tissues], the cadaverous and extracted tissues," and "bodies and the body parts of the socially marginal") are "transformed into technologies to aid in the intensification of vitality for other human beings" (19). Biovalue exists whenever human bodily material—organs, tissues, and whole cadavers—is "instrumentalized" in ways that make that human material useful in "science, industry, medicine, agriculture, and other arenas of technical culture" (33). Waldby originally introduced this concept in relation to biotechnologically rich environments in which experts translate the human body into informatic code or extract from it viable human tissue for biomedical research or donation, thus enabling it to circulate in what Waldby and Mitchell (2006) call "tissue economies" (31). The biovalue of stem cells or human blood is not only dependent upon its instrumentalization (being made into a useful technology) but also upon its physical separation from what we take to be the living body. By separating these tissues from bodies and "setting up certain kinds of hierarchies" (Waldby 2000, 19) among them, medical science seeks to "change their productivity along specific lines" (Waldby 2002, 310). This separation from the body and enactment of biovalue usually involves those bodies "at the margins of life or death," those that are "nearly dead or not-quite-alive" (Waldby and Squier 2003, 28).

Though Waldby (2002) identifies this process as taking place at "the level of the cellular or molecular fragment" and not at the level of the "macro-anatomical system" (310), I find that cadaveric bodies of the gross lab are also invested with biovalue through their instrumentalization and perceptual separation from the social and cultural body of living personhood. In the gross lab, biovalue is expressed in the cadaveric body as participants come to recognize the descriptive and relational evidence of the anatomical body, a formation of concepts imposed on and enacted by way of former living humans. The cadaver becomes the object of anatomy and the primarily text on which this anatomy is written. Through the development of trained vision, bodies of living participants become tools for enacting this anatomical body invested with biovalue. Or as James, a TA in the prosection course, said, "You need to touch things [like nerves, arteries, and veins], feel it, to know it." Through haptic and visual interactions with the world of objects, the living bodies form a type of intercorporeal technology that, through a particular type of embodied rhetorical action (one that is both physical and perceptual), transforms cadavers by anatomizing them—literally cutting into them and rendering them as anatomy. Participants' focus on cadaveric biovalue is made possible by the encomiastic discourses and procedures that exalt dissection.

A dental student, Roy, described the process in this way: "You realize that it is just tissue, and it is just a learning tool. And I haven't really thought about it much since then." The human body is valuable as a scientific specimen of anatomical knowledge, but understanding the body in this way does not entirely strip the body of its humanity; rather, it sensitizes participants to the cadaver's biovalue: "I was always fully aware that it is just a bunch of matter that made up the human body, and it was here for our learning" (Erin, undergraduate TA). For Erin, and others, the cadaver becomes a tool for understanding the component parts that "made up" (or constitute) the human. Though the body's personhood is downplayed, the biovalue, and thus the larger human relevance of the body, is emphasized. After all, biovalue occurs (or perhaps comes into view) when science instrumentalizes one body for the biomedical benefits of other bodies, in this case, doctors in training and their future patients. However, this focus on the body in parts and its biovalue inevitably reduces the body to its components:

> You just really start to see things in pieces rather than just, you know, this person, as a whole. And, and you are just so focused on the structures and everything else, then you are not paying attention to the body as a whole. (Samuel, medical student)

This rendering of the body as the body in parts is a consequence of anatomical vision. In other words, viewing bodies as science encourages participants to focus on the body's anatomical structures. Such a view increases participants' focus on the body as a collection of complex structures with descriptive and relational evidence. For many students and TAs, the act of dissecting bodies, of transforming them into corporeal objects of anatomical knowledge, encourages participants to view bodies as science and to hyperfocus on the body in parts. This perspective is encouraged by the lab's epideictic discourses and procedures.

However, it is incorrect to assume that the instrumentalization of the body and the focus on its biovalue empties cadavers of their more individualized personhood. Biovalue contains the same but-still structure of the cadaveric body. As the instrumentalization and vitalization of bodily material on the margins of life, biovalue's uniqueness and usefulness is rooted in its biology, in its biological being. Science can biotechnologically infuse human fetal tissue with biovalue because it is human tissue extracted from some human being. The nonindividualized humanity is present all along. The same is true in the gross lab; focused distance may, as I have shown, downplay the personhood of the cadaver, but it cannot completely remove all traces of humanity. The cadaver is valued because it is a once-living human. Some students even acknowledged their cadaver's humanity, either the nonindividualized humanity or the human agency of the former person who donated his or her body. Some TAs were proud of the way they could view the cadaver as an object of anatomy while acknowledging its humanity:

> I feel pretty comfortable working with the cadavers. There still are some things that get to me a little bit, that are, I feel like, a little bit disturbing or—But that's almost good in a way because then you know you are not completely desensitized to the entire thing. (Kimberly, undergraduate TA)

Kimberly's comfort with cadavers does not preclude moments when she feels a little "sensitized" (her words) to the fact that she is working on and with dead people.

A male medical student beautifully articulates a thought held by many TAs in the dissection course: a wish to remember the cadaver's personhood and even seek out traces of what that person's life might have been like:

> I start to think of it as more as, like, "This is a person. Maybe this is what her walk was like; maybe this is what their stature was like." And [I] really kind of get to know and start working with the hands. Or you start to see a mole on the skin. You really start to get this visual picture of what the person was like when they were walking around, you know. And I think you start to really look forward to finding the different aspects of what that person is like, internally and externally. And I think that helps a lot. (Brad, medical student)

For Brad, understanding the anatomical structures and biology of the body does not remove its humanity. He recasts his cadaver's humanity in bodily terms; his statements suggest an understanding of "living" bodily existence and embodied experience that can be read in and onto the cadaver's body. His sentiments are typical of students in the dissection course and TAs in the prosection course, who (during dissection) often reference the donor histories printed on cadaver tanks that listed the donor's ailments and diseases. In both cases, the cadaver becomes not only an instantiation of anatomical knowledge (a general anatomical body) but also an individual human being (a specific physical body). Their trained perspective that inevitably downplays individual personhood is balanced here by empathy with the former living human.

Conclusion

To conclude, I turn to the words of an undergraduate TA, describing how students get used to working with cadavers:

> I think it is just a matter of perspective. Um, when you start out, you see it more as like a dead human body, and that can be kind of upsetting. But later on you go to see it more as, you know, a specimen. The person's consented to giving their body and stuff like that, so just use it as a tool for learning. I remember when I took [the prosection course] I didn't actually touch a cadaver until the last few weeks of class. [He chuckles.] And, you know, I see some of the people doing the exact same thing, you know; they don't want to touch anything. Those are the ones who have trouble learning anatomy too, often. But, you know, as a TA, I kind of got thrown into it, having to dissect things. You know the person's dead, they don't—You know they knew what they were doing, and they didn't mind it. So I don't have a problem with that. I think it is just a matter of perspective. (Ned, prosection course TA)

Ned describes the anxiety and unease many students experience during their initial encounters with the cadaver, which is, first and foremost, a "dead human body." As they begin to work on and with these bodies, students often quickly move through this reaction. By dissecting, observing, and demonstrating, students come to see these bodies as anatomical specimen, objects that present anatomical knowledge. Ned seems to make peace with cadaveric anatomy by keeping in mind that these former living people donated their bodies to the anatomy program in order to contribute to the training of future medical professionals. Knowledge of donor consent allows Ned to focus on the purpose of anatomical education and the specific tasks at hand: the use of former living bodies as ideal tools for learning the anatomical body of medical science. Ned seems to take up the encomiastic accounts of cadaveric utility on which the lab's practices are premised; that is, the best way to learn the anatomical body—the conceptual body composed of anatomical terms, structures, and positions—is to view, touch, and cut into cadaveric bodies. By physically interacting with the cadaver, students and teaching assistants learn to recognize anatomical knowledge on and in these complex objects as well as to communicate that knowledge.

Ned remembers his own experience as a student in the prosection course, confessing he did not interact physically with cadavers until the course nearly ended. His own discomfort prevented him from taking advantage of the cadaver's tactile evidence. Later, when he became a TA, Ned had to overcome this hesitation immediately because his position as a coinstructor required that he dissect. Although he seems sympathetic to students' hesitation to engage with cadavers, Ned believes that those students who struggle to learn anatomy are frequently the same ones who take a hands-off approach to the lab. Speaking as a TA, Ned seems to find a correlation between firsthand experience with cadavers and the acquisition of anatomical knowledge. By telling the story of his change of heart, Ned acknowledges the more troubling moments while also offering a lesson to current students. Subtly, perhaps, Ned's story instructs students on an important cultural value of anatomical education, namely the need to physically interact with the body and to perceptually overcome any emotional barrier. One way to learn anatomy and make peace with anatomical practices is to focus on the activities of anatomy and the anatomical usefulness of cadavers—their biovalue. Ned, however, returns to the process of anatomical donation, saying that these people knowingly willed their bodies for the purpose of dissection and anatomical study. Ned's acknowledgment of this gift and, by implication, his belief in its worthiness, allows him to accept the invasive practices of cadaveric anatomy. In chapter 7, I explore this rhetorical formation that eulogizes cadavers as gifts.

The epideictic discourses and procedures that praise the uniqueness and utility of cadaveric bodies socialize students into the perspectives deemed important to anatomy education. Yet they also offer Ned a way of making sense of his experiences, even a way of reconciling his hesitant feelings about cadavers. For Ned and other participants, this epideixis subtly instructs them on how to think and feel about cadavers and the work of the lab. Again, Sullivan (1993a) describes education as a form of epideictic rhetoric that uses acts of praise and blame to teach students "reasoning appropriate" to the profession and to inculcate in students "sentiments or emotions" deemed worthy. Through a process of habituation, students internalize the ethos of that professional culture when they have learned to reason and feel like members of that group (71). In the gross labs, this encomiastic discourse contributes to the unique form of detached concern participants develop through their experiences with cadavers. This epideixis also structures how they experience the lab by authorizing as credible certain ways of thinking and ways of feeling while discouraging the possibility of others. In this educational space, as in others, the perceptual experiences formed through embodied activity are constituted by the multimodal objects and authorized practices of that setting. These perceptual experiences are also gained through enactive perception, activities that constitute our encounters with the world and are codetermined by the rhetorical discourses, displays, and procedures that saturate that domain of practice.

In contexts of technical training and work, epideictic discourses and procedures play formative roles. By learning which objects, practices, and attitudes the community seems to value and which it does not, we learn to think, see, and feel like members of our discipline or profession. In TPC, as Sullivan (1991) suggests, the epideictic impulse that animates our education can be witnessed in the objects and qualities we celebrate, the practices we seek to avoid, and the narratives of struggle and legitimation we tell. This epideixis is transmitted not just by the words of teachers, trainers, or mentors, but also by the procedures those professionals set in place—the assignments, tasks, and possibilities for action our courses, workplaces, and professions establish. Through these forms of epideixis, instantiated in discourses, procedures, and displays, we encourage our own versions of focused distance and encomiastic attachments that train students to attend to some meanings by silencing others. For instance, when we instruct students and interns to create certain kinds of images and displays, we are teaching them to appreciate the ways of seeing and thinking those displays make possible. Frequently cited research in technical visual rhetoric has forced us to reconsider the perhaps inadvertently unethical ways we present data by relying on visual tricks (Allen 1996) or downplaying human tragedy (Dragga and Voss 2001). After all, the examples we teach and the readings and tasks we assign become mediating artifacts that students and novices use to structure their experiences of our discipline. The documents, displays, discourses, and objects we extol and the procedures we put in place operate as socializing forces that develop the trained vision of our future professionals.

Imagine if we taught our introductory TPC courses using only documents and displays involving issues of human rights, environmental initiatives, or technical, scientific, and workplace controversies. The samples and examples we would use to teach genres and conventions would allow us opportunities to teach TPC as a field and set of practices inextricably linked to questions of ethics, power, and justice. Though many teachers do something like this in hope of training more truthful communicators, all of us indoctrinate our students in some way whether we recognize it or not. As TPC teachers and trainers, we develop students' knowledge and know-how, their communicative and disciplinary expertise, their embodied skills and perceptual experiences. In the process, we show them how and what to feel about the objects and practices that compose the work we do.

7

ACKNOWLEDGING PERSONHOOD

Anatomical Donation and the Gift Analogy

"silent teacher"
body

The praise of dissection and cadaveric anatomy is not the only epideictic discourse to structure the experiences and practices of the gross lab. The entire process and philosophy of anatomical donation is a result of, and perhaps overdetermined by, a rhetoric of benevolence and altruism—what I call "the analogy of the gift." This epideictic formation, which characterizes anatomical donation as a gift-giving process, is deeply rooted in anatomy education, including the labs I studied and the larger profession. In summer 2013, weeks before the annual conference of the American Association of Clinical Anatomists (AACA), members on the listserv received an email concerning the organization's new guidelines for acknowledging human subjects:

> The Anatomical Services Committee (ASC) would like to encourage all presenters who use data and/or images from anatomical donors to acknowledge those donors during their presentation or on their poster. The incredible work that you do can often be directly attributed to the valuable gift of human whole-body or tissue donor(s) and in turn, the anatomical donations our programs receive can be attributed to the quality work—your work—that result from these gifts. Please help us encourage donations and support anatomy research and education by recognizing anatomical donors. The ASC offers the following language for this purpose, *"The authors wish to thank individuals who donate their bodies and tissues for the advancement of education and research,"* or invites you to craft a statement of your own. (Brandi Schmitt, AACA).

In this message, Schmitt, an AACA council member and cochair of the ASC committee, urges conference presenters to acknowledge the "valuable gift" of donation by formally recognizing the sources of their work (cadaveric donors) in order to encourage donation through this gesture of respect. Schmitt even suggests language presenters might use and, later in the email, reminds them that *Clinical Anatomy*, the official journal of AACA, now requires that authors, in their acknowledgments, "thank the donor cadavers used in their study."

Whether in the gross lab or the conference auditorium, this persuasive discourse constructs anatomical donation as a benevolent act and the cadaveric body as a valuable gift to scientific knowledge, anatomical training, and the future of medicine—a gift with biovalue that continues to imply the individual personhood of the giver. As with the encomiastic discourses and procedures, participants filter their experience with cadavers through this rhetorical formation of bodies as gifts and donors as gift givers. Frank, a TA in the dissection course, illustrates how this rhetoric of the gift works in tandem with the encomium of dissection: "There is no substitute for the real thing. It is just great that people are willing to give that kind of a gift to teachers of medicine and future physicians. I cannot image any substitute." Taking up these discourses, Frank highlights the novelty of the cadaveric body, while characterizing it as a gift. His words demonstrate the subtle ways the lab's two key epideictic discourses intertwine to support each other. However, unlike the focused distance encouraged by the praise of cadavers, which downplays their former personhood, the analogy of the gift acknowledges a measure of individuality by viewing cadavers as gifts bestowed by autonomous human agents. According to the cultural implications of the anatomical gift system, this epideictic formation carries with it the subtle obligation to reciprocate, to give back. This gift analogy influences participants' daily interactions with cadavers and their thoughts about self-donation.

In this chapter, I argue that this analogy encourages a kind of empathy in those participants persuaded by its force, empathy that sensitizes them, first, to cadavers as donated gifts and, second, to their own bodies as possible future cadavers. This analogy works by emphasizing the individual personhood of the once-living cadavers, specifically the human agency involved in a donor's decision to donate her body to medicine. Taken together, the praise of dissection and the analogy of the gift distinguish the clinical detachment of the gross lab from its conventional formation as an objectification that precludes emotional sensitivity and acknowledgment of the body's humanity. In the gross lab, clinical detachment sensitizes participants to the uneasy tension between the science and personhood of the body. The twin epideictic formations that sustain this tension between the body as object and subject structure the experiences of participants and offer the rhetorical frames through which participants view dissection, observation, and demonstration.

The Gift and the Obligation to Reciprocate

The persuasive force of this gift analogy invites participants to treat cadavers as precious and invaluable objects bestowed to them by donors and their families. Remember the instructor's words on the first day of the prosection course:

"There are 50 cadavers for the course. That means 50 people donated their bodies. We require you to show respect for the human material. The people donated to teach anatomy, and not to be a spectacle for someone off the street." This gift analogy functions as an epideixis that honors and eulogizes the dead, much like the epitaphios logos, or funeral oration, of ancient Greece. In her historical study of the epitaphios in Athens, Loraux ([1981] 2006) asserts that this highly celebrated speech type was "a political genre" and an institution of the democratic city (42). Through it, speakers espoused the necessity of certain virtues by extolling how the dead exemplified them; the epitaphios uncovered in words what was admirable and noble about the dead and, by extension, espoused the venerable characteristics of Athens and the Athenian public.

By making public, in Loraux's ([1981] 2006) words, "a certain idea that the city wishes to have of itself," the epitaphios was a political performance that sought to produce a particular type of audience: one mobilized to value and respond to the notion of themselves as a public, thus offering a potentially durable form of public memory (42). The "codified praise" of one dead Athenian "spilled over into generalized praise of Athens" (27) by way of a synecdochal elision between the one being praised and those listening to the praise. This epideictic display, in the words of Perelman and Olbrechts-Tyteca ([1958] 1969), strengthened the community's "disposition toward action" by encouraging them to adhere to the values lauded (50). The epitaphios enacts a vision of the future by reflecting on a particular view of the past, one instantiated in the laudable actions of the dead. In the gross lab, this perspective translates into an exaltation of the selflessness, nobility, and generosity of cadaveric donors. The persuasive and ontological force of this epideictic analogy invites participants to honor and emulate the gift, at the same time extolling anatomical education as a valuable professional and personal experience.

Describing cadavers as gifts resonates with Burke's ([1945] 1969) description of metaphor as a "device for seeing something in terms of something else" (503). In this case, dead bodies are seen as gifts to the living. I describe this rhetorical construction as an analogy because it functions as an analogy does; namely, it is a style of reasoning, an extended argument, and a tool for invention, driven and constrained by the comparison set in play. The second-century CE rhetorical handbook *Rhetorica ad Herennium* ([Cicero] 1954) defines analogy as one of the figures of emphasis that "leaves more to be suspected than has been actually asserted" (401). Analogies work, in other words, by inviting implications not often expressed directly by the speaker. More specifically, according to Lanham (1991), analogy is a form of reasoning or argument involving parallel cases (10). Research in TPC and the rhetoric of science demonstrates the significance of analogy in technical domains as well as its unavoidable constraints. Graves's (2005) ethnography of an experimental physics laboratory exemplifies how scientists use analogy "to conceptualize abstract scientific phenomena" (82). Specifically, physicists' use of analogy as a mode of invention changed how they thought about experimental data. As Graves and Graves (1998) illustrate, analogies shape how we perceive an object, but their use may have unintended consequences (391). This complication stems partially from the inherent constraints of analogous reasoning. As Little (2008) asserts, the structures and correspondences that analogies set up constrain communicators. Little (2000, 2008) demonstrates this condition by reading scientific and technical uses of analogy through the lens of Gentner's theory of structure. In Gentner's (1983) cognitive model, analogy is a relationship structure of one domain that can be applied to another with the goal of emphasizing connections and relations between objects under comparison. For the analogy to be successful, the components of the objects being compared must match or, at least, relate to one another. When analogies do work, they foster significant insights, particularly when they connect back to an audience's prior knowledge (Gentner, Brem, Ferguson, Markman, Levidow, Wolff, and Forbus 1997). Gibson (2008) has even advocated a careful use of analogy in TPC as a deliberate tool for inductive reasoning.

If we take the constraints and consequences of analogy and analogous reasoning seriously, then rhetorically constructing the cadaver as a gift inexorably implicates anatomy students in a complex gift economy, a system of giving and receiving structured by reciprocal obligations. Mauss ([1950] 2000) understands rituals for gifts and gifting as part of a larger framework involving nearly every aspect of culture. For Mauss, gift giving includes three obligations: to give a gift, to receive the gift that is given, and to reciprocate the gift by giving back (39). Giving entails "recognition" on the part of the receiver, which compels the recipient to acknowledge the giver. If, for example, a friend gives me a gift, according to Mauss's formulation, she implicitly calls on me to recognize not only the object given but also her role (and social capital) as the giver. According to the conventions of such interaction, I am compelled to some expression of gratitude, which Mauss considers part of my obligation to receive the gift (40). If I do not accept it, I risk not only causing a rift with my friend but also losing face (41). After all, accepting this gift carries with it a subsequent need to give back to my friend an object of equal or greater value—in cultural if not strictly monetary terms (42). However, the gift my friend gives is not "inert" or "inactive." Even when the gift is in my possession, there is a cultural understanding that the gift "still possesses something" of the giver (12). This attitude is manifest in the fact that the gift will always be an object my friend gave me, a representation of her tastes or her perception of mine, as well as a symbol of what she wished me to have. Thus, the point of origin remains part of the object itself: "If one gives things and returns them, it is because one is giving and returning 'respects'—we all say 'courtesies.' Yet it is also because by giving one is giving oneself" (46). For Mauss, this powerful and active remainder of the giver implied by and even located within the gift is an almost spiritual essence. This gift exchange binds the giver and receiver in bonds of mutual "respects" because the gift one gives and receives is a gift of the self (46). The obligation to receive and give back is a crucial component of this ritual and, I contend, the reasoning behind the labs' gift analogy. The need for reciprocity influences the rhetorical deployment and consequences of this analogy in anatomy education.

Cadaveric Donation and the History of the Gift

According to encomiastic accounts of dissection, anatomical vision is possible because people donate their bodies or the bodies of family members for medical use. In labs that rely on cadaveric specimen as the primary teaching tools, this sentiment becomes a practical reality; without cadavers, students have no bodies to dissect, observe, and demonstrate with. American medical schools have a long history of using cadavers to foster and train this vision, and the gift analogy is arguably just as old. Warner and Rizzolo (2006) describe the way medical professors used this analogy in nineteenth-century American medical schools as a way of instilling a sense of empathy, or at least sympathy, in medical students (404). In the contexts of nineteenth- and early twentieth-century medicine, some anatomical donors conceived donating their postmortem body to anatomy as an altruistic act, intended to help educate physicians on anatomy and sometimes the specifics of a disease or ailment which afflicted the donor (Garment, Lederer, Rogers, and Boult 2007, 1,002; Warner and Edmonson 2009, 14).

Even today, this gift analogy has both an aspirational and practical purpose: to encourage respect and donation. At times, however, rather than anatomical donation, theft of the dead made cadaveric supplies possible. The analogy, sometimes written into the language of laws, was influential in passing anatomy acts regulating the use of bodies for medical education. To curb grave robbing in the eighteenth century, in 1789, New York became the first state to allow the bodies of executed criminals to be used in dissections. Over the next 40 years, Connecticut, Massachusetts, and Maine enacted similar legislation. In 1813, Massachusetts passed America's first official anatomy act, which applied only to Boston (Sappol 2002, 123). In 1913, 100 years later, all states with medical schools—save for four Southern states—had laws allowing these schools to claim the bodies of the indigent poor. In 1944, Louisiana followed suit, as did Tennessee in 1947 (124).[1]

In the mid-twentieth century, anatomical donation in America became more formal and regularized in order to protect the living and once living from unauthorized takings and use. In 1968, a special committee of the National Conference of Commissioners on Uniform State Laws (NCCUSL), charged with suggesting legislation that might be adopted by and enforced across all states, drafted the Uniform Anatomical Gift Act (Goodwin 2006, 11). After three years of study, the committee created the 13-provision document that determined the scope of cadaveric donations and defined what body parts could be donated (Waldby and Mitchell 2006, 164). The 1968 Uniform Anatomical Gift Act (UAGA) also provided an acceptable and standardized way for hospitals, universities, research institutes, and tissue banks to procure bodies and other tissue (Goodwin 2006, 111).

Specifically, two means of donation were authorized by the 1968 UAGA: (1) a person wishing to donate could "predesignate," by way of a will or some other witnessed document, his or her body for medical use; and (2) a family member could donate a deceased loved one's remains "even over the preexisting objections of that deceased" (111). All 50 states adopted the UAGA.

Interestingly, the act was not a consequence of some controversy involving the misuse of cadavers; instead, the UAGA was as a direct response to the developing medical procedure of organ transplantation and legal questions stemming from recent transplant surgeries, such as Barnard's 1967 successful transplant of a human heart (Sadler, Sadler, and Stason 1968, 2,501). By providing an agreed-upon definition and procedure for what the UAGA called anatomical "gifts," the committee hoped the act would encourage among the general public more organ and tissue donation, not to mention lessen the trepidation among physicians who wished to use cadaveric specimens but feared legal liability (Waldby and Mitchell 2006, 164). According to Stason, chair of the NCCUSL special committee, and Sadler and Sadler, two consultants to the committee, the UAGA sought to accomplish these goals by addressing four crucial yet "frequently competing" interests: (1) the request of the "decedent" (in the words of the original act); (2) the wishes of his or her spouse and next of kin; (3) the demand for human tissue and bodies for educational, research, and therapeutic purposes, including tissue transplant operations; and (4) "the need of society to determine cause of death in certain circumstances" (Sadler, Sadler, and Stason 1968, 2,501).

The philosophical basis of the act, at least as expressed by Sadler, Sadler, and Stason (1968), was the agency and autonomy of the donor (who had to be at least 18 years old and "of sound mind"), even at the expense of the surviving family: "The Uniform Act is based on the belief that an individual should be allowed to control the disposition of his own body after death and that his wishes should not be frustrated by his next of kin" (Sadler, Sadler, and Stason 1968, 2,503). This original act, however, left key questions unanswered: Could a donor be financially compensated for this gift? Did coroners and medical examiners have the right to "retrieve" tissue or organs from bodies in their care? Could a family member's objections override a donor's wish to bequeath? Furthermore, if the original goal was to increase organ donation, should hospitals and specifically physicians be required to formally request organ and tissue donation from patients and their next of kin (Goodwin 2006, 115)?

To settle some of these issues and increase the donation rate (which the 1968 UAGA failed to do), a number of congressional hearings in the 1980s led to both a revised UAGA in 1987 and the National Organ Transplantation Act three years earlier (Goodwin 2006, 113; Waldby and Mitchell 2006, 164). The 1987 UAGA, which only 26 states enacted according to the Uniform Law Commission of the NCCUSL, made it a felony to buy or sell an organ for transplantation (Uniform Law Commission 2011). Yet the 1987 act allowed for the passage of "presumed consent laws," laws that permitted some human tissue supposedly abandoned during medical procedures to be retrieved by physicians and used for research purposes (Goodwin 2006, 120; see also Skloot 2006; Waldby and Mitchell 2006). The most recent UAGA, enacted in 2006, sought to simplify "the document of gift," the language that effectuates donation and strengthens the rights of one wishing not to donate by allowing individuals "to sign a refusal" that "bars others" (particularly next of kin) from donating his or her body in whole or in part (Uniform Law Commission 2011).

All three iterations of the UAGA are important documents that establish procedures for the lawful donation of cadaveric bodies and tissue. The original 1968 act was perhaps most important because it "created the power" to donate one's body, or parts of it, to an authorized recipient such as an anatomy program. Though the analogy did not originate with the UAGA, the 1968 act did officially recognize and solidify anatomical donation as a "gift" and sought to regularize and protocolize the process of giving. By providing a legally sanctioned way for a decedent to become a "donor," the act's word for someone "who makes a gift of all or part of his [sic] body," the 1968 UAGA operates as a kind of legal and rhetorical rendering practice that allows the transformation of a corpse into an anatomical cadaver—a dead body into a gift. The transition of the body from former person to cadaveric specimen is visible in the gross lab, specifically in the ways participants come to view and interact with cadaveric bodies. Similar to the 1968 UAGA, this laboratory transformation is structured through authorized practices and supported by a rhetorical formation that constitutes these bodies as simultaneously decedents, cadavers, and gifts (Uniform Law Commission 2011).

Anatomical donation, like giving blood, establishes what Titmuss ([1970] 1997) terms "social ties of indebtedness between fellow citizens," ties that create "the condition for the maintenance of community between strangers" (309). Anthropologists and sociologists have written extensively about the ways we often understand organ and tissue donation as "giving the gift of life" (Fox and Swazey 1992; Lock 2002; Sharp 2006). The "seductive metaphor," as Lock (2002) calls it, renders donation "an act of charity" that is given to people in desperate need (317). As charity, these acts of benevolence imply the giver's "personal choice," a characterization that helps people reconcile the use of a person's body for the aid of another and squares with the belief in the right of self-determination. Remember Ned's words at the end of chapter 6: "The person's consented to giving their body." The implication of personal choice also obligates others to reciprocate. In organ and tissue donation, Fox and Swazey (1992) contend that this impossible obligation to reciprocate represents the "tyranny of the gift," the idea that gifts as precious as organs can never be fully reciprocated by the recipient (39). This special privileging of "gifts of life" or "anatomical gifts" functions as a posthuman version of the spiritual essence that Mauss ([1950] 2000) describes as inhabiting the gift (12).

Cadaveric Donation: Acknowledging Personhood

In whole-body donation, the gift becomes a biological object infused with biovalue that makes the body useful for medical research and education. The biovalue of this former person stems in part from the bequest process that articulates the dead body as a contribution to the future of health care. According to the bequest director, Patrick, and assistant director, Gillian, a large percentage of anatomical donors to this particular program are former medical professionals or long-term patients who struggled with cancer or other terminal diseases. Arguably, many people who give anatomical gifts do so because, as participants in the larger practices of medicine, they feel the obligation of the gift. For former doctors and nurses, donation is another way of dedicating themselves to medicine. For those suffering long-term illnesses, donation is a way to give back to the health-care system in return for what they have received. Though of course people are not literally obliged to donate their bodies, I suggest that the gift of themselves, this transformation of the body into a biovaluable gift, echoes Mauss's ([1950] 2000) formation of the three responsibilities of gift giving: to give, to receive, and to give back. They received a gift from medical science, and so they give themselves back to medical science. Students, TAs, and instructors of the gross anatomy lab participate in this gift economy, specifically in its obligation to reciprocate, in three interconnected ways: (1) by honoring the cadavers as gifts, (2) by memorializing the dead who gave, and (3) by deciding to become an anatomical gift themselves. These three gestures of obligation and the analogy that supports them infuse cadavers with a sense of personhood. Specifically, these eulogistic expressions of respect emphasize the former person's agency in making this altruistic decision.

Honoring the Gift

In the gross anatomy courses, instructors not only expose students to this gift analogy but also encourage them to reciprocate by respecting cadaveric bodies and taking advantage of the unique learning opportunity they provide. On the first day of both courses, Patrick, the bequest director, explains the donation process and the purpose of the bequest program: to procure anatomical donations for the cadaver-based anatomy courses. He cautions students to be considerate of donors while in the labs and when they leave. Moreover, he asks them to carefully consider how they talk about the lab experience, because "what you say about this may come across differently to different people. It is the subjective comment that you need to think about before you make it." Of course, he adds, "To talk about the topic, the subject, the science of anatomy, is not a problem." It is the comments that are "potentially disrespectful" that may be a problem. Patrick is asking the students to remember that the cadavers were living people who willed their remains to the lab. Thus, to speak crudely of dissection is to speak crudely of the donors and the choice they made. Of course, such talk is also detrimental to the reputation of the anatomy lab and cadaveric anatomy in general.

Patrick's exhortation encourages students to adopt the rhetorical discourse of gift giving and mutual reciprocity by reminding them to respect the donors and the donation process. The first reciprocal contribution that students offer donors, then, is to value the gifts by respecting donors' need for privacy, discretion, and anonymity. Instructors encourage students to not voice the details of anatomical dissection to those outside the community that might not be party to (and thus persuaded by) the gift analogy. The supposed goal of this selective speech is to protect the sanctity of the gift. Those who do not participate in the privilege of cadaveric anatomy might not fully understand how this gift functions and thus might misinterpret the frank speech of students who may or may not be disrespecting the program, the cadaver, and donor wishes.

Patrick provided the students a specific instance to emphasize his point:

> An example I point out is when two students were on the bus and a woman was sitting in front of them. And she called me, 30 years ago when this happened, and said, "If that's what you do I want no part of it." . . . And she shared with me that two students were talking about their anatomy experience on the bus. And she wasn't offended by the science of anatomy, but she was very much offended and disappointed by their lack of respect for the cadavers, the family, the medical school, the bequest program, even for her, not knowing she was a donor. (Patrick, bequest director)

The living donor was not "offended by the science of anatomy," which involves the dissection and evisceration of the body. What outraged the donor and led her to call the director was the fact that this loud conversation, disclosing the intimate details of what would someday happen to her body, took place on a public bus. Though no one on the bus knew she was a donor, the public discussion of what might be a private matter was, perhaps understandably, upsetting. Details of dissections complicate the analogy of the gift by emphasizing how students reciprocate that donation, namely by cutting into that gift. The economy of the anatomical gift, though characterized by a discourse of altruism, benevolence, and mutual responsibility, is complicated by the destructive practices of dissection. The students perhaps had no idea a donor could be a living person, one standing in earshot. The director tells this story to caution students and remind them of, in his words, their "great privilege" and their need to honor "the gift that these people have made to them and to their future."

All the instructors and bequest staff I interviewed seemed to understand the practical benefits of the analogy, namely ensuring a well-run lab free of controversy and unethical behavior. But each of those same instructors and members of the bequest staff seemed sincere in their conviction of the necessity of cadaveric dissection and the sanctity of the cadaveric gift. Their conviction reassures students and TAs of the importance of anatomical gifts. At the same time, it induces all participants to take up this rhetorical formation. For instance, most participants appreciated the bequest director's involvement, specifically his presentation at the beginning of the course:

> Then the bequest director came and talked to us at our orientation, which was really cool just to hear about the process of how people decided to donate, because I didn't really know anything about that. So it was like, O. K., you know, these people were well informed, and I feel like the bequest program really, you know, it looks out for its donors because it takes care of them, you know. They're doing it for this purpose. (Stacy, medical student)

By explaining the donation process, including how it works, what families and donors are told, and how students should view these bodies as a privilege, the director helps students understand the larger implications of the gift analogy. That is, the epideictic formation that characterizes the cadaver as a donor inevitably asserts the agency of the former person who gave their body for anatomical study and invites participants to honor this gift by giving their time, serious attention, and respect to the cadavers.

Students and TAs take up the gift analogy by honoring the cadaver and accepting the obligation to make good anatomical use of it. Note the words of Abby, a TA in the dissection course:

> You have this body, and it was this person's, you know, their wish, you know, to have us learn from them. And you are not honoring that wish if you do not take every opportunity to learn all that you can. And knowing that you always did then, I think, keeps you going if you are ever tripped up about it. At least for me, I cannot ever remember really being upset about it by any means.

Abby adopts the analogy of the gift and embraces the obligation of that gift. That is, accepting the cadaver as a gift means accepting the need to take full advantage of the learning opportunities that gift provides. For Abby, this formation of cadaveric donation as gift giving, with its implications of donor choice, allows her to work so closely with cadavers without getting "tripped up" or bothered by the invasive practices involved in dissection, demonstration, and observation. However, Abby goes on to acknowledge that her friends outside the lab do not understand her level of comfort. To explain to them how she approaches the troubling aspects of the course, she returns to the implications of the gift:

> It was their wish that they wanted us to learn from them, and it was our responsibility as anatomy students in this course to honor that wish. And it was a huge deal for them to dedicate their body to science, you know.

And so you have to respect it. So I think that keeps other people going too, because you really want to get all you can and not be wasteful.

Here, making peace with dissection depends on acknowledging and remembering the cadaveric body as a former person who made the decision "to dedicate their body to science."

As Abby's final words bear witness, this discourse of altruism is not without its complications, the greatest of which perhaps is that students are required to slowly and methodically eviscerate this gift they are asked to respect. One student, Daniel, attempts to explain the tension and ambiguity of the gift as both person and object:

> You realize that it is a teaching tool, and you respect it and that sort of thing. And that is kind of the feeling that I got. You know, you want to have respect for it, but you kind of have to realize that respect doesn't mean being timid during dissection. I think that is what I think I have learned kind of throughout the whole time. I mean, there [are] some times you have to use some blunt force to remove for—to get at and to find the anatomy. (Daniel, medical student)

To respect the donor and to honor the gift, participants must completely dissect the body, which can be a forceful, physically difficult activity. Rather than resolve this ambivalence, I contend that the epideictic discourses and procedures of the lab consistently encourage a formulation of cadavers as simultaneously specimen and former people. Students and TAs must come to grips with this ambivalence by remembering the individual personhood of the cadaver.

Students, TAs, and instructors frequently acknowledge the donor's choice whenever they reference anatomical donation. Participants in both courses used the knowledge that these bodies were gifted to the university as a type of touchstone to help them navigate the intellectual and emotional terrain of the course:

> It is more of a learning experience. And I think they know that when they donate their bodies. I try to keep in mind why they donated, and stuff like that, that they knew and they wanted us to be able to work" (Miranda, medical student).

By remembering that the cadaver was once a person, a person who willed her body specifically for medical education, students and TAs seem to gain solace that helps them cope with particularly difficult or invasive procedures. This ability to home in on a cadaver's former self is sometimes fostered by the physical remainders of that personhood.

Unlike tissue or organ donation, anatomical donors do not lose all traces of their individual personhood. Though the bequest staff wash and embalm all cadavers, they do not remove or conceal markers of individually. Tattoos, scars, fingernail polish, and prostheses (such as catheters and penile pumps) remain visible on the cadavers. The staff members do cover the faces until participants need them for dissection or observation, but this concealment is primarily to preserve the tissue and allow students a chance to acclimatize to cadavers before being confronted with the faces of the dead. A few participants did express feelings like one student's remark, "Oh, my God. You know, this looks like my grandma." However, many students describe over and again a wish to remember the importance of the cadaveric gift, one that implied an emotional engagement: "It's a human being, a person. It's a fellow human." This dissection student's statement illustrates the emotional complexity of witnessing the cadaver's ambivalent personhood.

This tension is present in the language that students and TAs use to describe the bodies. Participants learn to deploy anatomical terms and discourse, and the persuasive forces of the lab's epideictic formations shape their perspectives and attitudes toward the anatomical processes and objects. But participants also use language to make sense of, come to grips with, and express the experience of anatomical training, specifically their own intertwining relationships with cadavers. One legend of the anatomy lab experience is that students give their cadavers names; however, most participants I interviewed and observed either did not name their cadavers or decided not to share this information with me. Though one group referred to their cadaver as "Grandma" (because she was an elderly women when she died), most students seemed either uninterested in or uneasy with naming. As one female medical student put it, "They already have a name. I just don't know it." When I asked if he and his team had named their cadaver, one male medical student, Javier, stated, "It is not a dog or a car. It's not a pet. It's a person." Javier's use of the pronoun "it" makes his visible dismissal of my question ironic, yet it also bears witness to the complexity of the cadaveric body as simultaneously former person and current object. Perhaps because of this tension, instructors and TAs allow students to use whatever naming devices seem natural and comfortable; as long as all participants exhibit respect for cadavers, the pronouns and names used are of little consequence. Encouraged by instructors, students and TAs come to accept these cadaveric bodies as both educational tools (objects) and educational guides (instructors) in a way that seeks to reconcile the biology and personhood of these cadaveric gifts. One way instructors and staff do this is by encouraging students and TAs to honor not just the gift but also the donor's act of giving.

Memorializing the Giver

The program I studied uses most of the donations to educate medical, dental, mortuary science, and physical therapy students as well as undergraduates in the health professions. On occasion, and under the supervision of anatomy program staff, medical device engineers or practicing surgeons use cadavers to try out devices or surgical implants. Typically after 18 months in the custody of the program, cadavers are prepared for final interment. Interment involves cremating and returning the ashes to the family members, who may have held an empty-casket funeral for their loved one months before. When families do not wish to claim the body, participants lay the cremated remains to rest in a special area of a local cemetery. Each year in November, the anatomy program, through the bequest office, invites the family members to a memorial service designed to honor those who bequeathed their bodies and to thank the families who ultimately authorized that donation. At these memorials, first-year medical and dental students who worked with the bodies of that year's donors organize an evening of speeches, poetry, music, and dance. By planning, performing, and helping finance the event, these future physicians, dentists, and other profession- als "give back" to those, both living and dead, who made their anatomical educa- tion possible. Held in one of the university's largest concert halls, this formal, catered event provides anatomy students, faculty, and deans across the university a chance to honor and eulogize the sacrifice made by complete strangers.

As part of the tradition of these annual anatomical memorial services, students offer a gift to the families whose loved ones donated their bodies. Instructors and students describe these gifts as "small gestures" or "lasting objects" intended to symbolize the students' and the entire program's appreciation for the anatomi- cal gift. At the service I attended, medical and dental students presented a richly symbolic gift to each family present: the naming of a star using the International Star Registry. Students gave each family a certificate of authenticity and a celes- tial map pointing out Doctor Silens, Latin for "silent teacher," the star students named to memorialize the donors, those they understand to be their first and perhaps best teachers (see figure 7.1).

In Mauss's ([1950] 2000) terms, the students are fulfilling their cultural and personal obligation to give back by honoring both families and donors. Naming a star after the donors offers a permanent reminder first of the donor and their original gift and second of the students and their reciprocal gift.

By presenting the gift in the form of a celestial map, students share one manifestation of their anatomical vision with the donor families. These astro- nomical displays require the same visual rhetorical practices as anatomical atlas and the anatomical body; you find constellations by orienting yourself physi- cally and visually and then searching out neighboring celestial landmarks (see figure 7.2).

Figure 7.2 identifies the location of Doctor Silens in relation to Ursa Major. This constellation, commonly referred to as the Big Dipper, is emphasized on the map with connecting lines that outline the star formation. The circle (red in the original) to the right of Ursa Major is Doctor Silens. Just as students learn to observe and demonstrate the anatomical body using landmarks and relational values, the family members can locate this celestial object using a similar set of

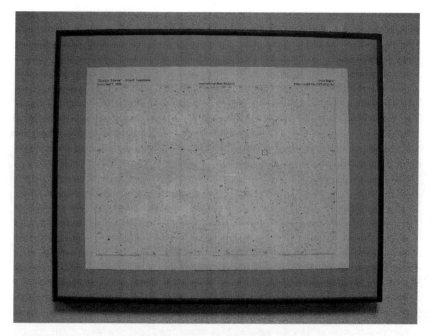

FIGURE 7.1 Celestial map of Doctor Silens given to the donor families.

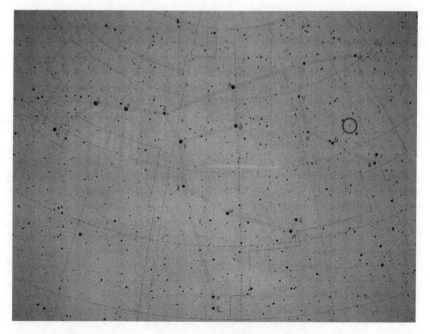

FIGURE 7.2 Detail of the celestial map identifying Doctor Silens.

demonstrational practices. Using similar processes encouraged by the enacted affordances of *Netter*, the donor's loved ones can use this display to make sense of the densely clustered night sky.

By calling the star "silent teacher," students directly engaged the idea of the cadaver as the source of anatomy education, anatomical vision, and their budding expertise. Even before the memorial service, I heard students repeatedly describing cadavers as their "first teachers." Though a few students did refer to the cadavers as their "first patients," most students and TAs overwhelming preferred the metaphor of "teacher." As one female TA in the dissection course mentioned, "We are not going to be doing this [dissection] to our patients, not even as surgeons." In a survey of medical students, Bohl, Bosch, and Hildebrandt (2011) found that 84% of the 125 responding students also preferred the notion of the "body as first teacher" as opposed to the "body as first patient." In line with my findings, they suggest that this metaphor might facilitate students' wish for a "closer personal relationship with donors," thus balancing "clinical detachment and empathy" (209). Bohl, Bosch, and Hildebrandt also suggest that anatomy programs such as the one in my study might benefit from first introducing the teacher metaphor and then later "transition[ing]" into the patient metaphor (212). These anatomists suggest that a more overt use of this rhetorical expression, which functions as an encomiastic articulation of the cadaver, might help facilitate a more empathetic formation of clinical detachment. This recognition of the power of rhetoric to shape action, attitudes, and beliefs is understandable in an academic discipline and medical profession so imbued with epideictic discourses and procedures.

One male TA's argument against the cadaver-as-patient metaphor speaks to the tension involved in cadaveric anatomy's use of rhetorical discourses to memorialize the cadaver as a noble gift:

> Especially, I think, the fact that people aren't alive adds another aspect to the initial difficulties that people have to get used to. I think when people are still living, and you were working so hard to kind of get them better or whatever, then I think it is easier to deal with it all. (Malcolm, prosection course TA)

Malcolm suggests that the physical presence of cadavers—the dead bodies of former medical patients (in one way or another)—contradicts the reason most students become physicians and dentists: to help improve and even save human lives. The fact that their efforts on the cadaver will not benefit it underscores the complexity of dissecting one body in hopes of later restoring another. Yet the gift analogy of the gross lab entails much more; it means accepting gifted bodies and giving back by cutting into them, taking seriously the opportunities of the lab, respecting cadavers, and finally dedicating oneself to medicine. This gift economy might also mean giving one's own body for anatomy education.

Deciding to Give

In the labs, the cadaver becomes invested with biovalue that renders the body a tool for education and research, an object now made beneficial to budding medical practitioners. In the case of cadavers, the biovalue does not nullify the personhood of the donor, though it does privilege certain aspects of the donor's identity and history. Take the medical histories posted on each cadaver tank. By highlighting only the body's ailments, the histories inevitably offer a pathologized personhood; however, the instructors and bequest staff do not espouse a pathologized view of the body. As I have mentioned previously, the courses encourage students to view cadavers as materializations of the anatomical body first and foremost, thus forgoing diagnostic interpretations until they have learned anatomy. As purveyors of the gift analogy, instructors and staff willingly allow moments when a cadaver's former personhood emerges, particularly instances that make apparent the person's decision to donate.

For instance, one female donor asked that a letter she wrote to her future dissectors be included with her body. William, the instructor of the prosection course, read aloud this letter, redacted of identifying markers, to the TAs who prepared her body. Then he posted it on a wall in the lab so that students might read it. In the letter, the donor tells of her life and the ways that struggles with cancer gave her a deeper respect for doctors and a wish to "give something back" (her words) in exchange for the excellent medical care she received. Her letter articulates her obligation to reciprocate, one matched by William, the anatomy instructor in charge. When I asked William why he read aloud and posted the letter, he reminded me that it was the donor's wish and that the program was dedicated to honoring such requests. He added that he wanted students and TAs to share in that moment, to understand "the sacrifice this person made for their education," and to understand the important "privilege" of working with whole-body cadavers.

Many students and TAs I interviewed contemplated the same decision made by "the lady with the letter," as she was frequently called: to become donors themselves. To understand how students and TAs view the donation process, I asked all interview participants if they had considered donating their own bodies to an anatomical gift program. These responses reveal their perceptions of and even personal investment in cadaveric donation. They also bring to light the contradictions involved in developing empathetic connections with cadaveric bodies. Though nearly all course instructors said that they were donors or planned to donate (should their loved ones approve), few students and TAs in either course answered definitively, and most answered with some degree of ambivalence or equivocation. Yet not one had ruled out the possibility of becoming a donor. Taken together, their responses fit into two broad categories: (1) those who replied predominantly "no," and (2) those who replied predominantly "yes" (see table 7.1 and table 7.2).

For students and TAs who answered predominantly "no," their primary reasons suggest an emotional identification of cadavers as potential sources of

TABLE 7.1 Predominantly "no" responses to the self-donation question.

The Body on Display: "But also, I guess, I just don't have any interest really in having my body dissected and looked at. And I don't think it's disrespectful, what is done to the cadavers, but I do feel like it is something I wouldn't want to have my body go through." (Sandra, prosection course TA)

Family's Concerns: "And I think probably, when I die it's best for everyone involved, my family, loved ones, if there is sort of just a service, and it is dealt with and laid to rest." (Keith, dissection course TA)

Too Much Information about Dissection: "I think that almost, um, ignorance of knowing what's going on would kind of set someone up better for donating their body here, or like to any anatomy research. . . . I just think that you know too much when you, when you TA here, you work here, um, about what happens with them." (Joan, dissection course TA)

Concerns about Public Knowledge: "But I don't know if the general population knows, and if they would be comfortable with exactly what we did." (Naomi, dissection course student)

TABLE 7.2 Predominantly "yes" responses to the self-donation question.

Cadaveric Body as an Anatomical Tribute or Sacrifice: "Yeah, I mean, I have so much respect for the people that can do that. I mean, that is amazing, you know, because it is such a sacrifice. Well, not a sacrifice. But it is for the families, you know. They'll not have that person anymore. I think for the family and the person it is a huge sacrifice." (Jeanette, prosection course TA)

Cadaveric Body as Process of Medical Socialization: "I had a heart condition, and I have already had surgery on it. So I would be so willing to donate my heart and my body, to say, 'Here. Figure out new ways to fix this before it becomes a problem.' Or I would even donate to CSI teams and be on a body farm, and let them figure out how things, you know, decompose." (Rebecca, prosection course student)

Transformation via the Course: "Regarding my body as a cadaver, I have considered it now. I've considered it. I haven't [makes a motion with his hand that implies signing his name] so to say, delegated it. I have considered it, due to this course. This course has influenced me a lot in many ways, many aspects of my life. And donating my— donating my body to science is one of the ways it has influenced me. I'm considering it now." (Sonny, dissection course TA)

selfhood and a reluctance to assume the position occupied by the cadavers. In other words, they "could not see," or did not want to imagine, themselves as a cadaver. To illustrate, I use a tabular arrangement to display interview responses that express the main themes of their answers (see table 7.1).

The families' wishes and their own wishes for them to remain a corpse and not become a cadaver underscore the responses in these first two groupings—the body on display and each family's concerns. First, many students and TAs could not see themselves, metaphorically speaking, as a cadaveric body because they could not image themselves as a naked and dissected body on display. Here they seem to be both projecting themselves onto the cadaveric body and projecting the cadaveric body onto their own embodied selfhood, and it is this mutual mapping that makes them uncomfortable. Even while expressing their discomfort about donation, these students were careful to reiterate an esteem for the process, insisting that it is not "disrespectful" (Sandra's words). They also worried what their loved ones would think of them gifting their bodies. Wishing to avoid distressing those they loved and to provide their family a chance to mourn, many participants thought they would probably opt out of donation.

Interestingly, the third and fourth themes of chiefly "no" responses almost seem to undermine the analogy of the gift by casting dissection in an unseemly light. For example, many TAs in particular felt that they knew too much about the process to want to participant as a donor, again implying a wish not to identify themselves and their bodies with this process. Similarly, a few participants wondered how the general public would feel about dissection if they could, as one medical student put it, "see it for themselves." In this set of reasons, participants are deciding to not donate (leaning more toward a "no" decision) based not on the process of dissection but on the ways the outside world perceives that process. I asked one of the medical students to "tell me more" about his discussion of this point, and he replied that he was not sure he wanted to his own body until he could hear exactly what he would be told by the bequest staff. In other words, he wanted to know how transparent the system was before he decided to contribute. When I further asked him if he had not already contributed by taking the course, he paused and answered: "I guess I am conflicted. I guess I don't want [the cadaver] to be me someday." I posit that behind these reasons to refrain from donating lays a perception of the self as alive and intact that stands in contradiction to the cadaver as dead and in parts. All these respondents who answered mainly "no," however, stated their wish to be organ donors if possible. Thus, I would argue they understand and are persuaded by the gift analogy, though they wish to make only a partial rather than whole contribution of their bodies. One respondent even described his preference for organ donation because it was "a gift of life."

Those who answered predominantly "yes" to whole-body donation openly and explicitly identified themselves as a potential cadaver because of their wish to contribute to anatomy education and medical practice by offering up their bodies as a tribute and even a sacrifice (see table 7.2).

Those more inclined to become a donor often expressed a respect for the cadaveric body that they used in class—or "my donor" (as one of them put it)—and the process of cadaveric gift giving. In the first response, Jeanette views giving her body to an anatomy program as a sacrifice she might make because she believes in the process of cadaver-based anatomy education. In the second passage, Rebecca not only communicates respect for the process but also intimates that she was already caught up in this gift-giving economy. Because of a heart condition that almost killed her, she had already benefited from medical care, so donation offered her a chance to give back by giving herself. Without a moment's hesitation, she states a seemingly sincere request not only to be part of an anatomy lab but even a forensics lab if it helped professionals "figure out new ways to fix" her heart condition before it develops. The last passage, which expresses a common sentiment among the TAs, illustrates the ways the gross lab experience transforms participants' views of the cadaveric body and the ways participants take up the epideictic discourses. Though this male TA had not officially become a donor, he had reconsidered what he described as his previous "absolutely-not stance" on self-donation. His experiences as a student and TA and his socialization in the anatomy labs allowed him to learn anatomy and become a teacher of anatomy and also led him to reconsider his position on anatomical donation.

Through the experience of anatomy education—the development of anatomical vision and exposure to the persuasive force of the analogy of the gift—participants learn to view the physical body as the anatomical body and to perceive cadavers as anatomical gifts. Some students and TAs, however, learn to appreciate how the cadaver becomes a possible self by coming to terms with their own position as a possible cadaver. Those students and TAs most open to whole-body donation, I suggest, went beyond establishing an empathetic connection with these "first teachers." By either definitely or tentatively committing themselves to this gift-giving economy, these participants began to feel with and not for the cadavers. This sense of feeling along with the cadavers and the donors is fostered in part by the epideictic rhetorics that shape the actions, attitudes, and beliefs of gross anatomy students, TAs, and instructors.

Conclusion

For many participants, to engage in anatomical education is to be persuaded by the epideictic discourses and formations that shape the practice. Through the analogy of the gift, students and TAs become aware of cadaveric bodies as gifts made by other human beings who wished to give back to medicine. The knowledge of this gift allows many participants to make peace with the lab's practices and induces them to contribute to this gift-giving ritual. By acknowledging or at least recognizing the possible convergence between their bodies and the ones on the table, living participants come to understand their overlap with the dead. What the living and the once living share is a reciprocal relationship to anatomical giving that implies not only interplay but the possibly of reversal: the recipient becomes the giver. This crisscrossing can encourage an empathetic relation in students who make peace with the dead body on the table by making peace with the dead body they will become. The wish to become a donor, even the ambivalence involved in taking that decision seriously, encourages participants to reconcile the anonymity of the cadaver with the individually of the donor.

This reconciliation is a form of empathetic engagement with cadavers and the once living persons whose traces the bodies still carry. This mode of detached concern, which sensitizes participants to appreciate the body as both biology and personhood, is a consequence of anatomical vision, the ways of seeing, thinking, and being structured by embodied practices and rhetorical formations.

One female medical student beautifully expresses this sensitization as she contemplated bequeathing her body to an anatomy program. (To further ensure anonymity, I have changed some of the details in this interview excerpt.)

> My boyfriend's cousin was in a horrible car accident when she was 20, which was four years ago, and she had—there were so many people who wanted us to donate her organs. And her father kind of, who was the legal guardian, he decided that they were going to continue her on life support and feeding tubes, and now she is a permanent vegetable in like hospice care or a good Samaritan home.
>
> And she is just, she just doesn't do anything, and I think it's terrible that— like I haven't seen her in three years, because I kind of said goodbye already. And, I mean, I think it is kind of equivalent to Terry Schiavo, a little bit, but she [the young woman] has obviously a little more cognitive function than Terri Schiavo, but it is pretty minimal. And, I mean, it is pretty bad.
>
> But I think that people fear death too much. If I'm dead, yeah, yeah, give my body to train more physicians and for education. So I think I would. I have considered it. I am not an official you know [chuckles] real donor yet, but yes I want to. (Stacy, medical student)

Stacy explains how she came to her decision to potentially donate through the unfortunate situation of the young woman who, in Stacy's estimation, is being made to live primarily because her family members (perhaps understandably) are not ready to deal with her death. Stacy's comparison to the Terry Schiavo case expresses her view of the medical, political, and philosophical implications of severe brain damage and life-support systems. Stacy implies that this woman's body could be put to better use as a cadaveric specimen because the life she is living now does not seem like much of a life to Stacy. Stacy uses these cases to claim that people often "fear death too much," a mindset that prevents them from dealing with the death of a loved one. Though not officially a donor, which she describes with a chuckle as "you know, a real donor," Stacy considers donation because she does not fear death and by implication is not unsettled by the idea of being a cadaver. In addition, she wants her death and dead body to be made meaningful as a tool for training health-care professionals. According to the gift analogy, Stacy seems ready to fulfill her implied obligation by giving herself to future medical students.

Stacy's comments approach the philosophical and religious by intimating that cadaveric dissection is an option because her conscious self will no longer reside in the dead body. Her discussion describes how one human body, also without consciousness, is being kept alive for another person's benefit. She would rather not be kept alive in this way, but instead she wants her death to benefit other people though anatomical study. Thus, the physical body can be made useful for medical science as a cadaveric body, or it can be made useless by medical science as a body in a vegetative state. Here we see the double-edged blade of medical science and biopower, the power to prolong life or allow death. Her words also imply that the usefulness, or biovalue, of the body-in-death potentially trumps a more personal attachment to that same body.

I would argue that Stacy exemplifies both clinical detachment and empathy. First, she views the dead body through an objectifying anatomical vision that emphasizes its instrumental biovalue over its former personhood. Second, she seems to empathize with the young woman's current situation and the dead body she could be. After emotionally and intellectually putting herself in the woman's place, Stacy is able to say that perhaps she would like for her dead body to become a cadaveric one. What is not addressed, of course, are the wishes of the young woman who may or may not have wanted to be kept alive and who may or may not have wanted to become a cadaver.

Examining how participants conceptualize cadaveric anatomy and whole-body anatomical donation offers another glimpse into the persuasive and onto-logical effects of epideictic discourse as well as its role in the development of dispositional tendencies that shape attitudes, actions, and practices. Anatomists are often well aware of the persuasive force of these rhetorical formations. But it is important to note that these discourses and practices exert a persuasive and ontological force on students, TAs, and instructors as well as potentially anyone habitually exposed to these practices, anyone who comes to identify with the gift analogy. Through my time in the anatomy program, I have undoubtedly been socialized to perceive anatomical donation as altruistic. In fact, I, like many par-ticipants I observed and interviewed, wrestle with the decision to donate my own body for anatomy education. Though, like most students I interviewed, I have not officially become a donor, I consider it a serious possibility. I have imagined myself as a cadaver on a table, and I am not particularly disturbed by the pros-pect. Perhaps I contemplate becoming a cadaveric donor because of my obser-vations of and nascent socialization into the lab's embodied rhetorical actions. Or perhaps I am persuaded by the rhetoric that shapes not only the gift-giving processes and the cultural meaning attributed to those gifts but also the larger structures and practices of anatomy education that make anatomical bodies of everyone. Maybe I entertain this option out of some personal sense of indebted-ness; after all, my work and this book are what they are because of those cadavers. It might be that I wish for my physical flesh to again intertwine with the phenom-enological flesh of these experiences and these practices in death as it did in life.

In the gross lab, the body is an object and a subject, an instantiation of techni-cal knowledge and a vehicle for lived experience. The body is all of these things due to the way participants engage with it and the discourses that shape those embodied engagements. In any domain of technical work (whether offices, labs, or classrooms), embodied practices and authorizing discourses equally structure the objects, documents, and displays that surround us and make enactive processes possible. These authorizing discourses take a number of forms and serve a multitude of purposes—from the technical language we learn, to the procedural instructions we follow, to the metaphors and analogies we use to make sense of the cultural and technical objects and actions of our environment.

In TPC, complex rhetorical, specifically epideictic, discourses and formations infuse our spaces of work and training and shape our experiences in the profession by offering us ideas and values to emulate or avoid as well as objects and displays to appreciate or dismiss. Gaining technical expertise in TPC requires developing the knowledge, skills, and habits of mind prized by the profession. But it also means learning to embody the values and attitudes of the community. We acquire these ways of seeing, knowing, and feeling through epideictic and apodeictic processes of enculturation and habituation, structured by the intertwining of bodies, objects, and discourses.

8

CONCLUSION

Embodied Rhetorical Action

Learning anatomy is a matter of perspective, conceptually and literally. Through repeated exposure to the multimodal displays and rhetorical discourses of the lab, participants learn to enact the anatomical body—the conceptual body of anatomical knowledge—in and on the physical human body. This enactment is specifically made possible through the embodied and rhetorical actions involved in demonstrating, observing, and dissecting the objects and bodies of the lab. These visual, haptic, and kinesthetic activities of seeing, touching, and cutting require the skilled use of multimodal displays and objects, as well as the anatomical and institutional discourses that structure and even authorize those embodied activities. As participants become socialized into these habituated practices that develop bodily and rhetorical skills, they adopt the perspective of an anatomist. They develop a trained perspective that the lab's apodeictic displays of proof and epideictic displays of communal values make possible. To develop anatomical expertise, that advanced amalgam of knowledge and know-how, participants must learn to see, think, and even feel as an anatomical expert. This perspective, which I call anatomical vision, is constructed through the embodied and rhetorical practices of anatomy education. By learning how to view, touch, move, use, and communicate with these anatomical displays, students and TAs develop the trained vision that coincides with expertise; this has been the major argument of *Rhetoric in the Flesh*.

In the process of making this argument, I have posited the physical human body as foundational to technical domains of practices. I have done so by putting forward a theoretical framework drawn from rhetorical theory and phenomenological theories of enactive cognition, a framework that articulates the way embodied practices and rhetorical displays mutually construct and constitute one another. In any technical domain—from classrooms, to research labs, hospitals, software companies, and construction sites—we develop trained vision through our rhetorical and embodied engagements with the objects, displays, discourses, and documents that constitute that domain of practice. This trained vision plays a crucial role in structuring and generating our thoughts, perceptions, and possibilities for action, and it does so by shaping the ways our bodies respond to our environment. Embodied rhetorical action is the name I give to this body-world coupling that we experience and enact through our skillful and rhetorical interactions with the world around us.

We are not exactly the objects and discourses we use, but we are materially and conceptually intertwined with them. As we move in the world, we learn to enact the world around us by fitting objects to our bodies and our bodies to objects. These objects, whether material documents or nonphysical discourses, take hold of us, and we them. Bodies, objects, and discourses mutually articulate each other through embodied rhetorical actions that give these objects their meaning. Embodied rhetorical action, drawing on the mechanisms involved in enactive cognition and the suasory force of rhetorical formations, is the way we learn to see and use objects. Habituation in a domain of practice allows, in Merleau-Ponty's words, "the things to pass into us as well as we into the things" (1968, 123). After all, ways of seeing the world are not just ways of knowing the world, but they are also ways of being in the world that make the seeing, the knowing, and the very objects of that world (including ourselves) possible in the first place.

Technical practices stem from this intertwining of multimodal objects and rhetorical discourses that coalesce into embodied rhetorical actions. These actions train vision by training bodies to respond to the objects and discourses in particular ways. Technical expertise, then, is not about having and not having; it is about seeing, doing, and being—using the body's capacities to develop and enact the knowledge, objects, and practices of that technical domain. Meaning making and communication are not disembodied processes confined to the brain, eyes, or head. Rather, they are activities we accomplish with and through our bodies. There can be no TPC—as a field of practice or an academic discipline—without the development of trained vision that we develop through socialization, habituated action, and moment-to-moment meaning-making processes. All of these activities depend on the body-object-environment assemblage that is embodied rhetorical action.

By asserting that all technical domains operate by way of embodied rhetorical action, I do not wish to imply that the setting-specific operations of an anatomy lab are common to all scientific, medical, and technical settings. The physical labor of the gross lab is markedly different from the labor involved in TPC workplaces or classrooms. Even still, anatomy, rhetorical education, and TPC have a great deal in common. Anatomy as a discipline and a profession is deeply rooted in a concern for education, for training students to recognize, understand, and appreciate the anatomical body's structure–function relationship. While anatomists conduct research, compete for grants, and publish research-based work in peer-reviewed journals like any other academic scientists, anatomists are also teachers. Formed in 1983, the AACA strives to develop "the science and art of clinical anatomy," to encourage "research and publication," and to "maintain high standards in the teaching of anatomy" (Dawson 1988, 237). This organization is a response to the decline in anatomy instruction in the United States and works to "strengthen ties between clinicians [mostly surgeons] and anatomists" (238).

Historically speaking, rhetoric is also a teaching field, or at least a field of study rooted in the rhetorical and philosophical training of others (Haskins 2004; Hawhee 2004; Walker 2011). In fact, what separates the field of rhetorical studies from being merely a specialized form of literary criticism is, among other things, rhetoric's disciplinary commitment to the training of speakers, writers and, increasingly, multimodal communicators (Walker 2011). Technical communication as well is a field with deep connections to pedagogy. Much of the research published in our professional journals is motivated, often explicitly, by training-related issues and research questions. Questions of training or education animate nearly all of our research, whatever the specific objects of study we choose. Inevitably, many of us write up our work in a form we hope will allow it to move outside the academy and address the concerns of practitioners—the writers, designers, developers, consultants, and trainers who make up the various subfields of TPC. As in anatomical and rhetorical education and practice, TPC participants learn through prolonged encounters with the objects and discourses of that domain, through embodied and rhetorical encounters that allow them to develop the skills, knowledge, habits, attitudes, and values that result from trained vision.

Becoming a technical communicator, then, involves developing not only communicative skills and disciplinary knowledge, which cannot be easily separated, but also developing the trained vision and bodily skills required to perform the tasks of technical communicators. This development occurs through repeated exposure to the activities and settings where technical communication happens as well as repeated practice with the objects of those settings. Involving students in service-learning activities and partnerships with for-profit and nonprofit industries, for example, helps to accomplish such development. Like Tardy (2009) and others, I am convinced that students can learn to recognize, appreciate, and create genres of scientific, medical, and technical communication in a classroom. Tardy (2009) offers a persuasive approach to teaching genre knowledge that involves both practice and rhetorical awareness.

But performing TPC tasks will not make someone a technical communicator any more than dissecting bodies will make someone an anatomist. Entering a discipline or profession requires inculcation in the ways of knowing and being valued by that domain. That epideictic education comes from TPC instructors and trainers, as well as from a student's exposure to workplace settings. Internships, for example, might involve reflective components that require students to consider not only the documents and skills of that workplace, but also the embodied practices, the modes of reasoning, and the epideictic formations that structure that culture. Like Henry (2000), I am suggesting that we encourage our students to become ethnographers- and archaeologists-in-training so that they have tools to analyze and understand a workplace's culture and systems of acculturation. I am also suggesting they do the same kind of rhetorical analysis of our field so that they understand the epideixis and apodeixis structuring the workplaces, classrooms, discourses, displays, and documents of TPC.

This heightened awareness begins with the training of instructors and supervisors. Whether for our own TPC courses or writing-in-the-disciplines (WID) initiatives in other fields, we need to introduce novice teachers to the ways scientific, medical, and technical domains are constituted through embodied rhetorical actions that shape the vision and build the expertise of participants. To train new instructors in this way means raising our own awareness of the types of physical, embodied actions that scientists, engineers, doctors, and other technical workers perform. While a TPC or WID instructor's primary goal is never to know particle physics or civil engineering with the same expertise and vision as a physicist or engineer, the communication instructor must know the documents and displays these practitioners create, as well as the reasons for their creation and the roles they play in socializing participants and building their expertise. TPC and WID instructors can gains such knowledge by investigating the culture and research of these fields—that is, through exposure to the embodied practices, apodeictic displays, and epideictic impulses that infuse those fields.

Observations of technical research labs or workplaces would offer TPC and WID instructors and researchers insights into the embodied practices and multimodal displays that structure work in scientific and technical disciplines, work that students are both learning and learning to communicate. Those less interested in or unable to conduct observational explorations of technical domains in action can turn to a wealth of available print and online materials. Nearly every technical discipline, including our own, generates trade magazines and other practitioner-focused media. These sources are a wonderful place to explore the apodeictic and epideictic formations of a technical discipline without putting in hours of qualitative observations. Approaching expertise as a consequence of embodied rhetorical action, then, offers instructors of TPC, WID, and composition important disciplinary insights into the development of communicative skills as a form of trained perspective.

Similarly, an embodied rhetorical approach to expertise can also provide necessary political opportunities for TPC instructors and writing programs. All too often in WID collaborations, the role of writing and communication specialists is reduced to issues of style and editing; they are brought in primarily to "fix" students' grammar or teach them "formats" for writing. Both of these positions not only reinforce the dual framework that unhelpfully divides expertise into content and form (Geisler 1994), but they also fail to acknowledge the complex rhetorical, cognitive, and physical processes involved in all forms of technical work—from conducting the experiments to writing about them. I maintain that we should train TPC and WID teachers to appreciate a more complex formation of expertise that defies dichotomous configurations one based on embodied practices and rhetorical displays. By understanding how embodied rhetorical action forms expertise, TPC and WID instructors and trainers can teach communication—written, spoken, or multimodal—as knowledge-making, or knowledge-enacting, processes inextricably linked to disciplinary practices. These same instructors, as well as administrators, can better articulate the expertise we bring and encourage our students and disciplinary colleagues to question a simplistic division between content and form.

In settings of TPC training, including classrooms and workplaces, students and practitioners can benefit from a greater focus on the embodied and rhetorical discourses and procedures that make possible scientific, medical, and technical domains. We can develop in our students a more nuanced understanding of the ways epideixis and apodeixis structure the work and culture of these domains. In the process, we will be engaging them in the embodied activities and multimodal displays we use to conduct our work and to build our professional vision. Like the students of the gross lab, as TPC instructors and practitioners, we engage in embodied processes of enactive cognition that our own epideictic and apodeictic demonstrations codetermine. By helping students understand this, we can help them gain a greater capacity to recognize, observe, and evaluate the implicit and explicit forms of epideixis and apodeixis that surround them. Developing such capacities is, after all, what a rhetorical training is all about.

APPENDIX: DATA COLLECTION AND ANALYSIS

Research is an act of poiesis, of making. Conducting research involves learning to see by way of the embodied rhetorical actions necessary for a discipline, a method, or an approach. Inspired by the insights of Clifford (1990) and Geertz (1973), I view ethnography as an interpretative method that produces an insightful, useful, and perspectival truth founded in the participants of that social setting and the subjectivity of the researcher. Unlike traditional ethnography, however, I am specifically concerned with capturing how participants seek to make sense of their own methods and meaning-making practices. As such, my approach shares a great deal with the ethnomethodology of Garfinkel (1999) and Lynch (1997). Like them, I seek to shed light on how participants structure their everyday knowledge of the world, and I do so by investigating the "embodied, communicative performance" of participants engaged in practice (Lynch 2003, 6). I use this brief appendix to provide more information about my methods of data collection and analysis. Again, I conducted this research with the approval of the University of Minnesota's IRB (study #: 0608E90306) and the consent of the anatomy program. All names used are pseudonyms. Tables A.1 and A.2 offer a detailed account of my data collection.

Data Collection

TABLE A.1 Data collected from laboratory courses.

ANAT 600: Medical and Dental Gross Anatomy Course

1. Ethnographic Observations: In fall 2006, I conducted direct observations of 100% of the dissection-focused lectures, 100% of the dissection labs, and 90% of the preparation labs of ANAT 600.

(Continued)

2. Interviews: I conducted interviews with the 15 TAs of the course and 15 students of the course (7 dental students and 8 medical students). I also interviewed 3 of the 4 main instructors of the course. These interviews lasted 45 minutes on average.

3. Textual Materials: I collected the following texts: (1) the syllabus and weekly schedule, (2) the course textbooks, (3) the laboratory manual, (4) PowerPoint slides of the morning lectures, and (5) additional material posted on the course's content management site.

4. Photographic Documentation: Each week, instructors and TAs changed the lab's physical spaces to provide peripheral learning material to accompany each session's dissections. Each week, I took digital photographs to document these changing spaces. I photographed (1) the whiteboards, (2) the posters and signs, (3) the medical imaging equipment (such as microscopes), and (4) the tools and instruments used.

ANAT 302: Undergraduate Anatomy Laboratory

1. Ethnographic Observations: During spring 2007, I conducted direct observations of ANAT 302. I observed 40% of the morning lectures (ANAT 301), 95% of at least two laboratory sections per week (ANAT 303), and 85% of the Friday afternoon prep labs, in which TAs prepared prosections and spaces for the following week.

2. Interviews: I conducted interviews with 15 TAs of the course and 4 undergraduate students of the course. I also interviewed the instructor of the course, who also directs the undergraduate laboratory. These interviews lasted 42 minutes on average. My original plan was to interview 15 undergraduates from the prosection course, but I was unable to find enough volunteers. A number of potential reasons for this exist: my close relationship with the course instructor and many TAs, the nearly ten-year age gap between me and most undergraduates, and my status as a graduate instructor in another department. Though I repeatedly discussed the confidentiality of their participation, I am convinced many students saw me as someone working with the course.

3. Textual Materials: I collected the following texts: (1) the syllabus and weekly schedule, (2) the course textbooks, (3) the laboratory manual, (4) the course notes (created by the instructors), and (5) the course material posted on the course's content management site.

4. Photographic Documentation: As with ANAT 600, I took digital photographs each week of the changing lab spaces: (1) the whiteboards, (2) the posters and signs, (3) the medical imaging equipment, and (4) the tools and instruments used.

TABLE A.2 Data collected from the bequest program and the memorial service.

The Anatomical Bequest Program

1. Interviews: To understand how the anatomical bequest program works, namely how people donate their bodies and how these donation move from corpse (dead body) to cadaver (anatomical specimen), I interviewed all four staff of the bequest program; I conducted these 45-minute, interviews in spring 2007.

2. Textual Materials: The bequest program texts that I collected include the following: (1) forms and documents needed to become a donor; (2) promotional material and public documents of the bequest program (e.g., websites, brochures); (3) legal definitions of what constitutes the "highest level of kinship" necessary to authorize the donation of others; (4) documents on the history of the bequest program; and (5) statistical information on past donors.

Annual Anatomy Memorial Service

1. Ethnographic Observations: In November 2006, I observed and took notes on the annual memorial service.

2. Textual Materials: The memorial service is professionally video recorded, and the recording is transferred onto a DVD so that each donor family can receive one copy. The DVD is meant to memorialize the event in a lasting format. The director of the bequest program gave me permission to view and transcribe that year's DVD, particularly the speeches, readings, and the short video presentation that the students created.

Data Analysis

As guides to qualitative interviewing, I turned to Seidman (2006) and Kvale (1996), especially their discussions on creating open-ended questions, working with the politics of the field, ethically interviewing, interacting with participants, and coding interview data. I transcribed all interviews within 36 hours. After completing each transcript, I double-checked it against the audio versions at least once. In spring 2007, I began using voice-recognition software to speed up the transcription process. Using a program called Dragon Naturally Speaking, I listened to the interviews and spoke aloud what I heard in the recordings. In fall 2009, electronic files containing several interview transcriptions were corrupted due to a software error. In summer 2010, a graduate student, Hannah Rankin, retranscribed the audiotapes with voice-recognition software. In fall 2010, I confirmed the accuracy of her transcriptions.

In creating my field notes, I used a procedure Emerson, Fretz, and Shaw (1995) call the "participating-to-write" style, meaning my major role in the labs was as an observer who sought to record activity, and not a full-fledged participant. To help me make sense of these activities, I studied anatomy using Netter's *Atlas of Anatomy* at least three hours each week during each semester of fieldwork. Within 24 to 36 hours of each observation, I typed my handwritten notes into a more official form. These became the field notes that I used as data. In analyzing typed field notes and transcribed interviews, I turned to an analytical-coding method suggested by Emerson, Fretz, and Shaw (1995). This method of emergent coding, involving an overall reading and a line-by-line investigation of the entire set of data, can be divided into six (often recursive) steps: (1) careful and repeated reading, (2) interrogation, (3) open coding, (4) initial memo writing, (5) focused coding, and (6) writing cumulative memos and outlines (143).

I used observations, interviews, photographic documentation, and textual analysis as means of data triangulation—the use of more than one source of data to gain a perspective that a single source could not achieve. By triangulating data, theory, and methodology, I sought to collect a robust set of data and perform a more rigorous analysis (Denzin 1978). Obviously, however, all research methods, no matter how triangulated the design or careful the researcher, contain their own particular affordances and limitations. In the spring of 2007, I began using qualitative analysis software, NVivo 6, to keep track of my emerging codes and my large volume of data. Eventually, I began supplementing my use of this software with an admittedly old-fashioned method employing hard copies and colored markers. I turned back to manual analysis to complement my computer-based analysis because I found it easier to keep track of each interview participant's larger narratives.

My original analysis of the data as expressed in my dissertation did not reflect my interest in enactive cognition nor did it address my perspective on embodied rhetorical action. Both of these, as with my turn to apodeixis and epideixis, developed during my reexamination of data as I sought to convert my dissertation into a monograph. To make better sense of my data and my experience of the labs, I turned to these rhetorical and phenomenological theories because they best articulated the perspectival truth I sought to uncover. In the steps and stumbles involved in this project—including the pilot, original analysis, dissertation, reassessment and reanalysis, and finally this book—I have found that my ways of seeing, thinking, and being have been transformed by the embodied and rhetorical action required to complete this work. The same is true for any researcher: each of us enact meaning and perspective through the research methods that train our vision.

NOTES

Note to Chapter 2

1. My discussion of incorporation and extension seems to share a great deal with cognitive theories of extension. These theories hold that some cognitive processes are made possible by the tendency to mesh with objects of the physical environment. In these cases of situated practice, as the body interacts with objects in the world, the mind seems to extend into the human's surroundings, and objects function as bodily extensions (Clark 2011; Menary 2010a; Rowlands 2010). For critiques of this approach, see Adams and Aizawa (2008) and Rupert (2009).

Note to Chapter 3

1. When the vestibulocochlear nerve was known as the audio-vestibular nerve, and when medical education was largely made up of male professors and students, the popular mnemonic was "Oh, Oh, Oh To Touch And Feel A Girl's Vagina." I never heard this misogynist and heterosexist mnemonic mentioned in the lab by any instructor or TA; I only learned of it from a medical student who found it on *Wikipedia*.

Note to Chapter 7

1. Alabama and North Carolina were the other two states without such anatomy acts. Sappol (2002) suggests these states' large prison systems, "with disproportionately large populations of black prisoners," provided the needed supply in the form of dead inmates (329n6). For more on the use of African Americans as cadaveric specimen in American history, see Blakely and Harrington (1997) and Washington (2006).

REFERENCES

Adams, Frederick, and Kenneth Aizawa. *The Bounds of Cognition*. Malden, MA: Blackwell Publishing, 2008.

Allen, Nancy. "Ethics and Visual Rhetorics: Seeing Is Not Believing Anymore." *Technical Communication Quarterly* 5, no. 1 (1996): 87–105.

Anderson, Daniel. "The Low Bridge to High Benefits: Entry-Level Multimedia, Literacies, and Motivation." *Computers and Composition* 25, no. 1 (2008): 40–60.

Arsenault, Darin J., Laurence D. Smith, and Edith A. Beauchamp. "Visual Inscriptions in the Scientific Hierarchy." *Science Communication* 27, no. 3 (2006): 376–428.

Aristotle. *On Rhetoric: A Theory of Civic Discourse*. Translated by George A. Kennedy. 2nd ed. New York: Oxford University Press, 2007.

Ball, Cheryl E. "Show, Not Tell: The Value of New Media Scholarship." *Computers and Composition* 21, no. 4 (2004): 403–25.

Banister, John. *Anatomical Tables*. London, ca. 1580.

Barad, Karen. *Meeting the Universe Halfway: Quantum Physics and the Entanglement of Matter and Meaning*. Durham, NC: Duke University Press, 2007.

Bazerman, Charles. *Shaping Written Knowledge: The Genre and Activity of the Experimental Article in Science*. Madison, WI: University of Wisconsin Press, 1988.

Benedetti, Alessandro. *Historia corporis humani sive Anatomice*. Venice: Bernardino Guerraldo Vercellensi, 1502.

Benner, Patricia. *From Novice to Expert: Excellence and Power in Clinical Nursing Practice*. Menlo Park, CA: Addison-Wesley, 2001.

Berengario da Carpi, Jacopo. *Commentaria cum amplissimis additionibus super anatomiam Mundini . . .* Bologna: Heironymum de Benedictis, 1521.

Berg, Marc. "Problems and Promises of Protocols." *Social Science & Medicine* 44, no. 8 (1997): 1081–88.

Berger, John. "Understanding a Photograph." In *Classic Essays on Photography*, edited by Alan Trachtenberg, 291–94. New Haven, CT: Leete's Island Books, 1980.

Berkenkotter, Carol. *Patient Tales: Case Histories and the Uses of Narrative in Psychiatry*. Columbia, SC: University of South Carolina Press, 2008.

Bezemer, Jeff, and Gunther Kress. "Writing in Multimodal Texts: A Social Semiotic Account of Designs for Learning." *Written Communication* 25, no. 2 (2008): 166–95.

Blair, Carole. "Contemporary U.S. Memorial Sites as Exemplars of Rhetoric's Materiality." In *Rhetorical Bodies*, edited by Jack Selzer and Sharon Crowley, 16–57. Madison, WI: University of Wisconsin Press, 1999.

Blakely, Robert L., and Judith M. Harrington, eds. *Bones in the Basement: Postmortem Racism in Nineteenth-Century Medical Training.* Washington, DC: Smithsonian, 1997.

Block, Ned. Review of *Action in Perception*, by Alva Noë. *Journal of Philosophy* 102, no. 5 (2005): 259–72.

Böckers, Anja, Lucia Jerg-Bretzke, Christoph Lamp, Anke Brinkmann, Harald C. Traue, and Tobias M. Böckers. "The Gross Anatomy Course: An Analysis of Its Importance." *Anatomical Sciences Education* 3, no. 1 (2010): 3–11.

Bogost, Ian. *Persuasive Games: The Expressive Power of Videogames.* Cambridge, MA: MIT Press, 2007.

Bohl, Michael, Peter Bosch, and Sabine Hildebrandt. "Medical Students' Perceptions of the Body Donor as 'First Patient' or 'Teacher': A Pilot Study." *Anatomical Sciences Education* 4, no. 4 (2011): 208–13.

Borell, Merriley. "Training the Senses, Training the Mind." In *Medicine and the Five Senses*, edited by W. H. Bynum and Roy Porter, 244–61. Cambridge, UK: Cambridge University Press, 1993.

Bourdieu, Pierre. *The Logic of Practice.* Stanford, CA: Stanford University Press, 1980.

Bucchi, Massimiano, "Images of Science in the Classroom: Wall Charts and Science Education, 1850–1920." In *Visual Cultures of Science: Rethinking Representational Practices in Knowledge Building and Science Communication*, edited by Luc Pauwels, 90–119. Hanover, NH: Dartmouth College Press, 2006.

Burke, Kenneth. *A Grammar of Motives.* Berkeley, CA: University of California Press, 1969. First published 1945 by Prentice-Hall.

Burke, Kenneth. *A Rhetoric of Motives.* Berkeley, CA: University of California Press, 1969. First published 1950 by Prentice-Hall.

Burke, Kenneth. *Permanence and Change: An Anatomy of Purpose.* 3rd ed. Berkeley, CA: University of California Press, 1984. First published 1935 by New Republic.

California Science Center. "Summary of Ethical Review: *Body Worlds: An Anatomical Exhibition of Real Human Bodies.*" 2005. Accessed April 10, 2013, www.bodyworlds.com/Downloads/englisch/Media/Press%20Kit/BW_LA_SummaryofEthicalReview.pdf.

Campbell, John Angus. "The Invisible Rhetorician: Charles Darwin's 'Third-Party' Strategy." *Rhetorica* 7, no. 1 (1989): 55–85.

Carey, Christopher. "Epideictic Rhetoric." In *A Companion to Greek Rhetoric*, edited by Ian Worthington, 236–52. Malden, MA: Blackwell Publishing, 2007.

Carlino, Andrea. *Books of the Body: Anatomical Ritual and Renaissance Learning.* Translated by John Tedeschi and Anne C. Tedeschi. Chicago: University of Chicago Press, 1999. First published 1994 by Giulio Einaudi.

Carter, Michael. "Ways of Knowing, Doing, and Writing in the Disciplines." *College Composition and Communication* 58, no. 3 (2007): 385–418.

Carter, Michael F. "The Ritual Function of Epideictic Rhetoric: The Case of Socrates' Funeral Oration." *Rhetorica* 9, no. 3 (1991): 209–32.

Cartwright, Lisa. *Screening the Body: Tracing Medicine's Visual Culture.* Minneapolis: University of Minnesota Press, 1995.

Charmaz, Kathy, and Richard Mitchell. "Grounded Theory in Ethnography." In *Handbook of Ethnography*, edited by Paul Atkinson, Amanda Coffey, Sara Delamont, John Lofland, and Lyn Lofland, 160–74. Thousand Oaks, CA: Sage, 2001.

Charon, Rita. *Narrative Medicine: Honoring the Stories of Illness*. New York: Oxford University Press, 2006.

Chase, J. Richard. "The Classical Conception of Epideictic." *Quarterly Journal of Speech* 47, no. 3 (1961): 293–300.

Chemero, Anthony. *Radical Embodied Cognitive Science*. Cambridge, MA: MIT Press, 2009.

[Cicero]. *Rhetorica ad Herennium*. Translated by Harry Caplan. Loeb Classical Library series. Cambridge, MA: Harvard University Press, 1954.

Clark, Andy. *Supersizing the Mind: Embodiment, Action, and Cognitive Extension*. New York: Oxford University Press, 2011.

Clifford, James. "Notes on (Field)notes." In *Fieldnotes: The Making of Anthropology*, edited by Roger Sanjek, 47–70. Ithaca, NY: Cornell University Press, 1990.

Collins, Harry. *Tacit and Explicit Knowledge*. Chicago: University of Chicago Press, 2010.

Colombo, Matteo Realdo. *De re anatomica*. 2nd edition. Parisiis: In officina. Joannis Foucherii junioris, 1562.

Condit, Celeste Michelle. *The Meanings of the Gene: Public Debates about Human Heredity*. Madison, WI: University of Wisconsin Press, 1999.

Connolly, William E. *Neuropolitics: Thinking, Culture, Speed*. Minneapolis, MN: University of Minnesota Press, 2002.

Connors, Robert J. "The Rise of Technical Writing Instruction in America." *Journal of Technical Writing and Communication* 12, no. 4 (1982): 329–51.

Cunningham, Andrew. *The Anatomical Renaissance: The Resurrection of the Anatomical Projects of the Ancients*. Aldershot, UK: Scolar, 1997.

Cunningham, Andrew. *The Anatomist Anatomis'd: An Experimental Discipline in Enlightenment Europe*. London: Ashgate Press, 2010.

Daston, Lorraine, and Peter Galison. *Objectivity*. New York: Zone Books, 2007.

Dawson, David. "The American Association of Clinical Anatomists: The Beginnings and First Five Years." *Clinical Anatomy* 1, no. 4 (1988): 237–58.

De Jaegher, Hanne. "Social Understanding through Direct Perception? Yes, by Interacting." *Consciousness and Cognition* 18, no. 2 (2009): 535–42.

De Jaegher, Hanne, and Ezequiel A. Di Paolo. "Participatory Sense-Making: An Enactive Approach to Social Cognition." *Phenomenology and Cognitive Science* 6, no. 4 (2007): 485–507.

Denzin, Norman K. *The Research Act: A Theoretical Introduction to Sociological Methods*. 2nd ed. New York: McGraw-Hill, 1978.

Denzin, Norman K. "Triangulation." In *The Blackwell Encyclopedia of Sociology*, edited by George Ritzer, 5,075–80. Vol. 10, *ST–Z*. Malden, MA: Blackwell Publishing, 2007.

Descartes, René. "Meditations on First Philosophy." 1641. In *Philosophical Essays and Correspondence*, edited by Roger Agnew, 97–141. Indianapolis, IN: Hackett Publishing, 2000.

Dinsmore, Charles E., Steven Daugherty, and Howard J. Zeitz. "Student Responses to the Gross Anatomy Laboratory in a Medical Curriculum." *Clinical Anatomy* 14, no. 3 (2001): 231–36.

Donnell, Jeffrey. "Illustration and Language in Technical Communication." *Journal of Technical Writing and Communication* 35, no. 3 (2005): 239–71.

Dragga, Sam and Dan Voss. "Cruel Pies: The Inhumanity of Technical Illustrations." *Technical Communication* 48, no. 3 (2001): 265–74.

Drake, Richard L., Jennifer M. McBride, Nirusha Lachman, and Wojciech Pawlina. "Medical Education in the Anatomical Sciences: The Winds of Change Continue to Blow." *Anatomical Sciences Education* 2, no. 6 (2009): 253–59.

Dreyfus, Hubert L. "Merleau-Ponty and Recent Cognitive Science." In *The Cambridge Companion to Merleau-Ponty*, edited by Taylor Carman and Mark B. Hansen, 129–50. New York: Cambridge University Press, 2005.

Dreyfus, Hubert L., and Stuart E. Dreyfus. *Mind Over Machine: The Power of Human Intuition and Expertise in the Era of the Computer*. New York: Free Press, 1986.

Dumit, Joseph. *Picturing Personhood: Brain Scans and Biomedical Identity*. Princeton: Princeton University Press, 2004.

Ellis, Harold. "Teaching in the Dissecting Room." *Clinical Anatomy* 14, no. 2 (2001): 149–51.

Emerson, Robert M., Rachel I. Fretz, and Linda L. Shaw. *Writing Ethnographic Fieldnotes*. Chicago: University of Chicago Press, 1995.

Fahnestock, Jeanne. "Rhetoric in the Age of Cognitive Science." In *The Viability of the Rhetorical Tradition*, edited by Richard Graff, Arthur E. Walzer, and Janet M. Atwill, 159–79. Albany: State University of New York Press, 2005.

Foucault, Michel. *The Order of Things: An Archaeology of the Human Sciences*. New York: Vintage, 1994. First published 1970 by Random House.

Foucault, Michel. *The Birth of the Clinic: An Archaeology of Medical Perception*. Translated by A.M. Sheridan Smith. New York: Vintage, 1994. First published 1973 by Tavistock Publications.

Foucault, Michel. *Discipline and Punish: The Birth of the Prison*. Translated by Alan Sheridan. New York: Vintage, 1995. First published 1977 by Knopf Doubleday.

Foucault, Michel. *The History of Sexuality: An Introduction, Volume 1*. Translated by Robert Hurley. New York: Vintage Books, 1990. First published 1978 by Random House.

Fox, Renée C. *Essays in Medical Sociology*. New York: John Wiley and Sons, 1979.

Fox, Renée C. "Medical Uncertainty Revisited." In *Handbook of Social Studies in Health and Medicine*, edited by Gary L. Albrecht, Ray Fitzpatrick, and Susan C. Scrimshaw, 409–25. Thousand Oaks, CA: Sage, 2000.

Fox, Renée C. and Judith P. Swazey. *Spare Parts: Organ Replacement in American Society*. New York: Oxford University Press, 1992.

Fraiberg, Steven. "Composition 2.0: Toward a Multilingual and Multimodal Framework." *College Composition and Communication* 62, no. 1 (2010): 100–126.

Fredal, James. *Rhetorical Action in Ancient Athens: Persuasive Artistry from Solon to Demosthenes*. Carbondale, IL: Southern Illinois University Press, 2006.

French, Roger. *Dissection and Vivisection in the European Renaissance*. Brookfield, VT: Ashgate, 1999.

Fuchs, Thomas, and Hanne De Jaegher. "Enactive Intersubjectivity: Participatory Sense-Making and Mutual Incorporation." *Phenomenology and Cognitive Science* 8, no. 4 (2009): 465–86.

Fuller, Steve. "'The Rhetoric of Science': Double the Trouble?" In *Rhetorical Hermeneutics: Invention and Interpretation in the Age of Science*, edited by Alan G. Gross and William M. Keith, 279–98. Albany: State University of New York Press, 1997.

Galen. *De usu partium corporis humani*. Paris: Simonis Colinaei, 1528.

Galen. *De anatomicus administrationibus*. Basel: Andreas Cratander, 1531.

Garfinkel, Harold. *Studies in Ethnomethodology*. Malden, MA: Blackwell Publishers, 1999.

Garment, Ann, Susan Lederer, Naomi Rogers, and Lisa Boult. "Let the Dead Teach the Living: The Rise of Body Bequeathal in 20th-Century America." *Academic Medicine* 82, no. 10 (2007): 1,000–1,005.

Geertz, Clifford. "Thick Description: Toward an Interpretive Theory of Culture." In *The Interpretation of Cultures*, edited by Clifford Geertz, 3–30. New York: Basic Books, 1973.

Geisler, Cheryl. *Academic Literacy and the Nature of Expertise: Reading, Writing, and Knowing in Academic Philosophy*. Mahwah, NJ: Lawrence Erlbaum Associates, 1994.

Gentner, Dedre. "Structure-Mapping: A Theoretical Framework for Analogy." *Cognitive Science* 7, no. 2 (1983): 155–70.

Gentner, Dedre, Sarah Brem, Ronald W. Ferguson, Arthur B. Markman, Björn B. Levidow, Phillip Wolff, and Kenneth D. Forbus. "Analogical Reasoning and Conceptual Change: A Case Study of Johannes Kepler." *The Journal of the Learning Sciences* 6, no. 1 (1997): 3–40.

George, Diana. "From Analysis to Design: Visual Communication in Teaching Writing." *College Composition and Communication* 54, no. 1 (2002): 11–39.

Gibson, James. *The Ecological Approach to Visual Perception*. Hillsdale, New Jersey: Lawrence Erlbaum Associates, 1986.

Gibson, Keith. "Analogy in Scientific Argument." *Technical Communication Quarterly* 17, no. 2 (2008): 202–19.

Gillingwater, Thomas H. "The Importance of Exposure to Human Material in Anatomical Education: A Philosophical Perspective." *Anatomical Sciences Education* 1, no. 6 (2008): 264–66.

Gogalniceanu, Petrut, Hardi Madani, Parashevas Parasheva, and Ara Darzi. Letter to the Editor. "Anatomy Teaching in the 21st Century—Dead Cool or Dead Cold?" *Anatomical Sciences Education* 1, no. 3 (2008): 136–37.

Goldenberg, Maya J. "On Evidence and Evidence-Based Medicine: Lessons from the Philosophy of Science." *Social Science & Medicine* 62, no. 11 (2006): 2,621–32.

Good, Byron J., and Mary-Jo DelVecchio Good. "'Learning Medicine': The Construction of Medical Knowledge at Harvard Medicine School." In *Knowledge, Power, and Practice: The Anthropology of Medicine and Everyday Life*, edited by Shirley Lindenbaum and Margaret Lock, 81–107. Berkeley, CA: University of California Press, 1993.

Goodwin, Charles. "Professional Vision." *American Anthropologist* 96, no. 3 (1994): 606–33.

Goodwin, Charles. "Practices of Color Classification." *Mind, Culture, and Activity* 7, no. 1/2 (2000): 19–36.

Goodwin, Michelle. *Black Markets: The Supply and Demand of Body Parts*. New York: Cambridge University Press, 2006.

Grasseni, Cristina. "Skilled Vision. An Apprenticeship in Breeding Aesthetics." *Social Anthropology* 12, no. 1 (2004): 41–55.

Grasseni, Cristina. "Communities of Practice and Forms of Life: Towards a Rehabilitation of Vision?" In *Ways of Knowing: Anthropological Approaches to Crafting Experience and Knowledge*, edited by Mark Harris, 203–21. New York: Berghahn Books, 2007.

Graves, Heather. *Rhetoric In(to) Science: Style as Invention in Inquiry*. Cresskill, NJ: Hampton Press, 2005.

Graves, Heather Brodie, and Roger Graves. "Masters, Slaves, and Infant Mortality: Language Challenges for Technical Editing." *Technical Communication Quarterly* 7, no. 4 (1998): 389–414.

Gray, Henry. *Anatomy, Descriptive and Surgical.* London: John W. Parker and Son, 1858.

Griere, Ronald N., and Barton Moffatt. "Distributed Cognition: Where the Cognitive and the Social Merge." *Social Studies of Science* 33, no. 2 (2003): 301–10.

Gross, Alan. *The Rhetoric of Science.* Cambridge, MA: Harvard University Press, 1996.

Gross, Alan. "Medical Tables, Graphics and Photographs: How They Work." *Journal of Technical Writing and Communication* 37, no. 4 (2007): 419–33.

Gross, Alan, Joseph Harmon, and Michael Reidy. *Communicating Science: The Scientific Article from the 17th Century to the Present.* New York: Oxford University Press, 2002.

Gross, Daniel M. *The Secret History of Emotion: From Aristotle's Rhetoric to Modern Brain Science.* Chicago: University of Chicago Press, 2006.

Guerrini, Anita. "Alexander Monro Primus and the Moral Theatre of Anatomy." *The Eighteenth Century* 47, no. 1 (2006): 1–18.

Haas, Christina, and Stephen P. Witte. "Writing as Embodied Practice: The Case for Engineering Standards." *Journal of Business and Technical Communication* 15, no. 4 (2001): 413–57.

Habinek, Thomas. *Ancient Rhetoric and Oratory.* Malden, MA: Blackwell Publishing, 2005.

Hacking, Ian. *Representing and Intervening: Introductory Topics in the Philosophy of Natural Science.* New York: Cambridge University Press, 1983.

Hafferty, Frederic W. *Into the Valley: Death and the Socialization of Medical Students.* New Haven, CN: Yale University Press, 1991.

Hafferty, Frederick and R. Franks. "The Hidden Curriculum, Ethics Teaching, and the Structure of Medical Education." *Academic Medicine* 69, no. 11 (1994): 861–71.

Hanna, Stephen J. and Jane E. Freeston. Letter to the Editor. "Importance of Anatomy and Dissection: The Junior Doctor's Viewpoint." *Clinical Anatomy* 15, no. 5 (2002): 377–78.

Hanson, Norwood Russell. *Patterns of Discovery: An Inquiry into the Conceptual Foundations of Science.* New York: Cambridge University Press, 1958.

Haraway, Donna J. "Situated Knowledges: The Science Question in Feminism and the Privilege of Partial Perspective." *Feminist Studies* 14, no. 3 (1988): 575–99.

Harris, Randy Allen, and Sarah Tolmie. "Cognitive Allegory: An Introduction." *Metaphor and Symbol* 26, no. 2 (2011): 109–20.

Hartelius, Johanna E. *The Rhetoric of Expertise.* Lanham, MD: Lexington Books, 2010.

Haskins, Ekaterina. *Logos and Power in Isocrates and Aristotle.* Columbia, SC: University of South Carolina Press, 2004.

Hauser, Gerard A. "Aristotle on Epideictic: The Formation of Public Morality." *Rhetoric Society Quarterly* 29, no. 1 (1999): 5–23.

Hawhee, Debra. *Bodily Arts: Rhetoric and Athletics in Ancient Greece.* Austin, TX: University of Texas Press, 2004.

Henry, Jim. *Writing Workplace Cultures: An Archaeology of Professional Writing.* Carbondale, IL: Southern Illinois University Press, 2000.

Hildebrandt, Sabine. "How the Pernkopf Controversy Facilitated a Historical and Ethical Analysis of the Anatomical Sciences in Austria: A Recommendation for the Continued Use of the Pernkopf Atlas." *Clinical Anatomy* 19, no. 2 (2006): 91–100.

Hildebrandt, Sabine. "Capital Punishment and Anatomy: History and Ethics of an Ongoing Association." *Clinical Anatomy* 21, no. 1 (2008): 5–14.

Hildebrandt, Sabine. "Anatomy in the Third Reich: An Outline, Part 1. National Socialist Politics, Anatomical Institutions, and Anatomists." *Clinical Anatomy* 22, no. 8 (2009a): 883–93.

Hildebrandt, Sabine. "Anatomy in the Third Reich: An Outline, Part 2. Bodies for Anatomy and Related Medical Disciplines." *Clinical Anatomy* 22, no. 8 (2009b): 894–905.

Hildebrandt, Sabine. "Anatomy in the Third Reich: An Outline, Part 3. The Science and Ethics of Anatomy in National Socialist Germany and Postwar Consequences." *Clinical Anatomy* 22, no. 8 (2009c): 906–24.

Hildebrandt, Sabine. "Developing Empathy and Clinical Detachment During the Dissection Course in Gross Anatomy" [Letter to the editor]. *Anatomical Sciences Education* 3, no. 4 (2010): 216.

Hirschauer, Stefan. "The Manufacture of Bodies in Surgery." *Social Studies of Science* 21, no. 2 (1991): 279–319.

Hirschauer, Stefan. "Animated Corpses: Communicating with Post Mortals in an Anatomical Exhibition." *Body & Society* 12, no. 4 (2006): 25–52.

Holden, Luther. "Landmarks, Medical and Surgical." In *Anatomy, Descriptive and Surgical, by Henry Gray,* edited by William W. Keen, 1,035–40. 11th American ed. Philadelphia, PA: Lea Brothers & Company, 1887.

Horsley, Philomena. "Teaching the Anatomy of Death: A Dying Art?" *Medicine Studies* 2, no. 1 (2010): 1–19.

Hovde, Marjorie Rush. "Research Tactics for Constructing Perceptions of Subject Matter in Organizational Contexts: An Ethnographic Study of Technical Communicators." *Technical Communication Quarterly* 10, no. 1 (2001): 59–95.

Hull, Glynda A., and Mark Evan Nelson. "Locating the Semiotic Power of Multimodality." *Written Communication* 22, no. 2 (2005): 224–61.

Husserl, Edmund. *Phantasy, Image Consciousness, and Memory, 1898–1925.* Edited by J. B. Brough. Berlin: Springer, 2006.

Hutchins, Edwin. *Cognition in the Wild.* Cambridge, MA: MIT Press, 1995.

Hutchins, Edwin. "The Distributed Cognition Perspective on Human Interaction." In *Roots of Human Sociality: Culture, Cognition and Interaction,* edited by N. J. Enfield and Stephen C. Levinson, 375–98. New York: Berg, 2006.

Hutchins, Edwin. "Enaction, Imagination, and Insight." In *Enaction: Toward a New Paradigm for Cognitive Science,* edited by John Stewart, Olivier Gapenne, and Ezequiel A. Di Paolo, 425–50. Cambridge, MA: MIT Press, 2010.

Hutto, Daniel D., and Erik Myin. *Radicalizing Enactivism: Basic Minds Without Content.* Cambridge, MA: MIT Press, 2013.

Ihde, Don. *Bodies in Technology.* Minneapolis, MN: University of Minnesota Press, 2002.

Ingold, Tim. "From the Transmission of Representation to the Education of Attention." In *The Debated Mind: Evolutionary Psychology versus Ethnography,* edited by Harvey Whitehouse, 113–53. New York: Berg, 2001.

Institute for Plastination. "A Life in Science." Gunther von Hagens's *Body Worlds*: The Original Exhibition of Real Human Bodies. 2013. www.bodyworlds.com/en/gunther_von_hagens/life_in_science.html/.

Jack, Jordynn and L. Gregory Appelbaum. "'This is Your Brain on Rhetoric': Research Directions for Neurorhetorics." *Rhetoric Society Quarterly* 4, no. 5 (2010): 411–37.

Johnson, Carol Siri. "Prediscursive Technical Communication in the Early American Iron Industry." *Technical Communication Quarterly* 15, no. 2 (2006): 171–89.

Katz, Susan M. and Lee Odell. "Making the Implicit Explicit in Assessing Multimodal Composition: Continuing the Conversation." *Technical Communication Quarterly* 21, no. 5 (2012): 1–5.

Kennedy, George. *On Rhetoric: A Theory of Civic Discourse.* 2nd ed. New York: Oxford University Press, 2007.

Keränen, Lisa. "Hippocratic Oath as Epideictic Rhetoric: Reanimating Medicine's Past for Its Future." *Journal of Medical Humanities* 22, no 1 (2001): 55–68.

Klestinec, Cynthia. *Theaters of Anatomy: Students, Teachers, and Traditions of Dissection in Renaissance Venice.* Baltimore, MD: Johns Hopkins University Press, 2011.

Kosslyn, Stephen, William Thompson, and Giorgio Ganis. *The Case for Mental Imagery.* New York: Oxford University Press, 2006.

Kostelnick, Charles and Michael Hassett. *Shaping Knowledge: The Rhetoric of Visual Conventions.* Carbondale, IL: Southern Illinois University Press, 2003.

Kress, Gunther. "Gains and Losses: New Forms of Texts, Knowledge, and Learning." *Computers and Composition* 22, no. 1 (2005): 5–22.

Kress, Gunther and Theo van Leeuwen. *Reading Images: The Grammar of Visual Design.* New York: Routledge, 1996.

Kress, Gunther and Theo van Leeuwen. *Multimodal Discourse: The Modes and Media of Contemporary Communication.* New York: Arnold, 2001.

Kristeva, Julia. *Powers of Horror: An Essay on Abjection.* Translated by Leon S. Roudiez. New York: Columbia University Press, 1982.

Kuhn, Thomas S. *The Structure of Scientific Revolutions.* 3rd ed. Chicago: University of Chicago Press, 1962.

Kuppers, Petra. *The Scar of Visibility: Medical Performance and Contemporary Art.* Minneapolis, MN: University of Minnesota Press, 2007.

Kvale, Steinar. *InterViews: An Introduction to Qualitative Research Interviewing.* Thousand Oaks, CA: Sage, 1996.

Kynell-Hunt, Teresa and Gerald J. Savage, eds. *Power and Legitimacy in Technical Communication, Volume I: The Historical and Contemporary Struggles for Professional Status.* Amityville, NY: Baywood Publishing, 2003.

Kynell, Teresa, and Elizabeth Tebeaux. "The Association of Teachers of Technical Writing: The Emergence of Professional Identity." *Technical Communication Quarterly* 18, no. 2 (2009): 107–41.

Lakoff, George and Mark Johnson. *Metaphors We Live By.* Chicago: University of Chicago Press, 1980.

Lanham, Richard A. *A Handbook of Rhetorical Terms.* 2nd ed. Berkeley, CA: University of California Press, 1991.

Latour, Bruno. "Drawing Things Together." In *Representation in Scientific Practice,* edited by Michael Lynch and Steve Woolgar, 19–68. Cambridge, MA: MIT Press, 1990.

Latour, Bruno. "Netz-Works of Greek Deductions." Review of *The Shaping of Deduction in Greek Mathematics: A Study in Cognitive History,* by Reviel Netz. *Social Studies of Science* 38, no. 3 (2008): 441–59.

Latour, Bruno, and Steve Woolgar. *Laboratory Life: The Construction of Scientific Facts.* Princeton, NJ: Princeton University Press, 1986. First published 1979 by Sage.

Lawrence, Susan C. "Educating the Senses: Students, Teachers and Medical Rhetoric in Eighteenth-Century London." In *Medicine and the Five Senses,* edited by W. H. Bynum and Roy Porter, 154–78. Cambridge: Cambridge University Press, 1993.

Lay, Mary M. *The Rhetoric of Midwifery: Gender, Knowledge, and Power.* New Brunswick, NJ: Rutgers University Press, 2000.

Leder, Drew. "A Tale of Two Bodies: The Cartesian Corpse and the Lived Body." In *The Body in Medical Thought and Practice,* edited by Drew Leder, 17–35. Dordrecht, Netherlands: Kluwer Academic Publishers, 1992.

Lella, Joseph W., and Dorothy Pawluch. "Medical Students and the Cadaver in Social and Cultural Context." In *Biomedicine Examined*, edited by Margaret Lock and Deborah Gordon, 125–53. Boston, MA: Kluwer Academic Publishers, 1988.

Levinas, Emmanuel. *Ethics and Infinity: Conversations with Philippe Nemo*. Translated by Richard A. Cohen. Pittsburgh, PA: Duquesne University Press, 1985.

Lief, Harold I. and Renee C. Fox. "Training for 'Detached Concern' in Medical Students." In *The Psychological Basis of Medical Practice*, edited by Harold I. Lief, Victor Lief, and Nine R. Lief, 12–35. New York: Hoeber Medical Division, 1963.

Little, Joseph. "Analogy in Science: Where Do We Go From Here?" *Rhetoric Society Quarterly* 30, no. 1 (2000): 69–92.

Little, Joseph. "The Role of Analogy in George Gamow's Derivation of Drop Energy." *Technical Communication Quarterly* 17, no. 2 (2008): 220–38.

Liuzzi, Mondino de'. *Anathomia*. Padua: Petrus Maufer, ca. 1474.

Lock, Margaret. *Twice Dead: Organ Transplants and the Reinvention of Death*. Berkeley, CA: University of California Press, 2002.

Longo, Bernadette. *Spurious Coin: A History of Science, Management, and Technical Writing*. Albany: State University of New York Press, 2000.

Longo, Bernadette, and T. Kenny Fountain. "What Can History Teach Us about Technical Communication?" In *Solving Problems in Technical Communication*, edited by Johndan Johnson-Eilola and Stuart A. Selber, 165–87. Chicago: University of Chicago Press, 2013.

Loraux, Nicole. *The Invention of Athens: The Funeral Oration in the Classical City*. Translated by Alan Sheridan. New York: Zone Books, 2006. First published 1981 by Editiones de École des Hautes Etudes en Sciences Sociales.

Lynch, Michael. "Discipline and the Material Form of Images: An Analysis of Scientific Visibility." *Social Studies of Science* 15, no. 1 (1985): 37–66.

Lynch, Michael. *Scientific Practice and Ordinary Action: Ethnomethodology and Social Studies of Science*. New York: Cambridge University Press, 1997.

Lynch, Michael. "Ethnomethodology." In *Harold Garfinkel, Volume 1*, edited by Michael Lynch and Wes Sharrock, 5–7. Thousand Oaks, CA: Sage, 2003.

Maier, Carmen Daniela, Constance Kampf, and Peter Kastberg. "Multimodal Analysis: An Integrative Approach for Scientific Visualizing on the Web." *Journal of Technical Writing and Communication* 37, no. 4 (2007): 453–78.

Mailloux, Steven. *Disciplinary Identities: Rhetorical Paths of English, Speech, and Composition*. New York: Modern Language Association, 2006.

Martin, Emily. "The Egg and the Sperm: How Science Has Constructed a Romance Based on Stereotypical Male-Female Roles." *Signs: Journal of Women in Culture and Society* 16, no. 3 (1991): 485–501.

Mascagni, Paolo. *Anatomia Universale*. Florence: Batelli, 1833.

Massa, Niccolò. *Liber introductorius anatomiae sive dissectionis corporis humani*. Venice: Francisci Bindoni and Malphei Pasini, 1536.

Massumi, Brian. *Parables for the Virtual: Movement, Affect, Sensation*. Minneapolis, MN: University of Minnesota Press, 2002.

Mauss, Marcel. *The Gift: The Form and Reason for Exchange in Archaic Societies*. Translated by W. D. Halls. Foreword by Mary Douglas. New York: Norton, 2000. First published 1950 by Presses Universitaires de France.

McGarvey, Alice, Thomas Farrell, Ronán Conroy, Skantha Kandiah, and Stanley Monkhouse. "Dissection: A Positive Experience." *Clinical Anatomy* 14, no. 3 (2001): 227–30.

McKeon, Richard. *Rhetoric: Essays in Invention and Discovery.* Edited with an introduction by Mark Backman. Woodbridge, CT: Ox Bow Press, 1987.

McNeill, David. *Gesture and Thought.* Chicago: University of Chicago Press, 2005.

Menary, Richard, ed. *The Extended Mind.* Cambridge, MA: MIT Press, 2010a.

Menary, Richard. "Introduction to the Special Issue on 4E Cognition." *Phenomenology and Cognitive Sciences* 9, no. 4 (2010b): 459–63.

Merleau-Ponty, Maurice. *The Visible and the Invisible.* Edited by Claude Lefort. Translated by Alphonso Lingis. Evanston, IL: Northwestern University Press, 1968.

Merleau-Ponty, Maurice. *Phenomenology of Perception.* Translated by Colin Smith. New York: Routledge, 2005. First published 1945 by Gallimard.

Miller, Carolyn. "A Humanistic Rationale for Technical Writing." *College English* 40, no. 6 (1979): 610–17.

Mol, Annemarie. *The Body Multiple: Ontology in Medical Practice.* Durham, NC: Duke University Press, 2005.

Mol, Annemarie and John Law. "Embodied Action, Enacted Bodies: The Example of Hypoglycaemia." *Body & Society* 10, no. 2/3 (2004): 43–62.

Montgomery, Kathryn. *How Doctors Think: Clinical Judgment and the Practice of Medicine.* Oxford: Oxford University Press, 2006.

Moore, Charles, and C. Mackenzie Brown. "Experiencing *Body Worlds*: Voyeurism, Education or Enlightenment?" *Journal of Medical Humanities* 28, no. 4 (2007): 231–54.

Moore, Keith L., and Arthur F. Dalley. *Clinically Oriented Anatomy.* 5th ed. Philadelphia, PA: Lippincott, Williams, and Wilkins, 2005.

Netter, Frank H. *Atlas of Human Anatomy.* 4th ed. Philadelphia, PA: Saunders Elsevier, 2006.

Netz, Reviel. *The Shaping of Deduction in Greek Mathematics: A Study of Cognitive History.* New York: Cambridge University Press, 2003.

Nienkamp, Jean. *Internal Rhetorics: Toward a History and Theory of Self-Persuasion.* Carbondale, IL: Southern Illinois University Press, 2001.

Nienkamp, Jean. "Internal Rhetorics: Constituting Selves in Diaries and Beyond." In *Culture, Rhetoric, and the Vicissitudes of Life*, edited by Michael Carrithers, 18–33. New York: Berghahn Books, 2009.

Noë, Alva. *Action in Perception.* Cambridge, MA: MIT Press, 2006.

Noë, Alva. *Out of Our Heads: Why You Are Not Your Brain, and Other Lessons from the Biology of Consciousness.* New York: Hill and Wang, 2009.

Norris, Sigrid. *Analyzing Multimodal Interaction: A Methodological Framework.* New York: Routledge, 2004a.

Norris, Sigrid. "Multimodal Discourse Analysis: A Conceptual Framework." In *Discourse and Technology: Multimodal Discourse Analysis*, edited by Philip LeVine and Ron Scollon, 101–15. Washington, DC: Georgetown University Press, 2004b.

Nutton, Vivian. "Logic, Learning, and Experimental Medicine." *Science* 295, no. 5,556 (2002): 800–801.

Nutton, Vivian. *Ancient Medicine.* New York: Routledge, 2004.

Oakley, Todd. "The Human Rhetorical Potential." *Written Communication* 16, no. 1 (1999): 93–128.

Oakley, Todd. *From Attention to Meaning: Explorations in Semiotics, Linguistics, and Rhetoric.* New York: Peter Lang, 2009.

O'Malley, Charles D. *Andreas Vesalius of Brussels, 1514–1564.* Berkeley, CA: University of California Press, 1964.

Oravec, Christine. "'Observation' in Aristotle's Theory of Epideictic." *Philosophy & Rhetoric* 9, no. 3 (1976): 162–74.

O'Regan, J. Kevin, and Alva Noë. "A Sensorimotor Approach to Vision and Visual Consciousness." *Behavioral and Brain Sciences* 24, no. 5 (2001): 939–73.

Park, Katharine. *Secrets of Women: Gender, Generation, and the Origins of Human Dissection*. New York: Zone Books, 2006.

Pasveer, Bernike. "Representing or Mediating: A History and Philosophy of X-ray Images in Medicine. In *Visual Cultures of Science: Rethinking Representational Practices in Knowledge Building and Science Communication*, edited by Luc Pauwels, 41–62. Hanover, NH: Dartmouth College Press, 2006.

Perelman, Chaim, and Lucie Olbrechts-Tyteca. *The New Rhetoric: A Treatise in Argumentation*. Translated by John Wilkinson and Purcell Weaver. Norte Dame, IN: University of Norte Dame Press, 1969. First published 1958 by Presses Universitaires de France.

Pernkopf, Eduard. *Topographische Anatomie des Menschen*. Berlin and Wien: Urban and Schwarzenberg, 1937.

Potts, Liza. "Using Actor Network Theory to Trace and Improve Multimodal Communication Design." *Technical Communication Quarterly* 18, no. 3 (2009): 281–301.

Prelli, Lawrence J. *A Rhetoric of Science: Inventing Scientific Discourse*. Columbia, SC: University of South Carolina Press, 1989.

Prelli, Lawrence J. "Rhetorics of Display: An Introduction." In *Rhetorics of Display*, edited by Lawrence J. Prelli, 1–38. Columbia, SC: University of South Carolina Press, 2006.

Prentice, Rachel. "Drilling Surgeons: The Social Lessons of Embodied Surgical Learning." *Science, Technology, & Human Values* 32, no. 5 (2007): 534–53.

Prentice, Rachel. *Bodies in Formation: An Ethnography of Anatomy and Surgery Education*. Durham, NC: Duke University Press, 2013.

Propen, Amy. *Locating Visual-Material Rhetorics: The Map, The Mill & The GPS*. Anderson, SC: Parlor Press, 2012.

Pylyshyn, Zenon W. "Seeing, Acting, and Knowing." Commentary on "A Sensorimotor Approach to Vision and Visual Consciousness," by Kevin J. O'Regan and Alva Noë. *Behavioral and Brain Sciences* 24, no. 5 (2001): 999.

Pylyshyn, Zenon W. *Seeing and Visualizing: It's Not What You Think*. Cambridge, MA: MIT, 2003.

Quintilian. *Institutio oratoria*. Loeb Classical Library series. Translated by Donald A. Russell. Cambridge, MA: Harvard University Press, 2002.

Regan de Bere, Sam, and Alan Peterson. "Out of the Dissecting Room: News Media Portrayal of Human Anatomy Teaching and Research." *Social Science and Medicine* 63, no. 1 (2006): 76–88.

Regli, Susan Harkness. "Whose Ideas?: The Technical Writer's Expertise in *Inventio*." *Journal of Technical Writing and Communication* 29, no. 1 (1999): 31–40.

Reifler, Douglas R. "'I Actually Don't Mind the Bone Saw': Narratives of Gross Anatomy." *Literature and Medicine* 15, no. 2 (1996): 183–99.

Reiser, Stephen J. "Technology and the Use of the Senses in Twentieth-Century Medicine." In *Medicine and the Five Senses*, edited by W. H. Bynum and Roy Porter, 262–73. Cambridge, UK: Cambridge University Press, 1993.

Richardson, Ruth. *Death, Dissection, and the Destitute*. 2nd ed. Chicago: University of Chicago Press, 2000a.

Richardson, Ruth. "A Necessary Inhumanity." *Journal of Medical Ethics: Medical Humanities* 26, no. 2 (2000b): 104–6.

Richardson, Ruth. *The Making of Mr. Gray's Anatomy: Bodies, Books, Fortune, Fame.* New York: Oxford University Press, 2008.

Rivers, Nathan. "Future Convergences: Technical Communication Research as Cognitive Science." *Technical Communication Quarterly* 20, no. 4 (2011): 412–42.

Rosenfield, Lawrence W. "The Practical Celebration of Epideictic." In *Rhetoric in Transition: Studies in the Nature and Uses of Rhetoric,* edited by Eugene E. White, 131–55. University Park, PA: Pennsylvania State University Press, 1980.

Rosse, Cornelius. "Anatomy Atlases." *Clinical Anatomy* 12, no. 4 (1999): 293–99.

Rowlands, Mark. *The New Science of Mind: From Extended Mind to Embodied Phenomenology.* Cambridge, MA: MIT Press, 2010.

Rupert, Robert D. *Cognitive Systems and the Extended Mind.* New York: Oxford University Press, 2009.

Ryle, Gilbert. *The Concept of Mind.* Chicago, IL: University of Chicago Press, 1949.

Sadler, Alfred M., Jr., Blair L. Sadler, and E. Blythe Stason. "The Uniform Anatomical Gift Act: A Model for Reform." *Journal of the American Medical Association* 206, no. 11 (1968): 2,501–6.

Sanders, John T. "Affordances: An Ecological Approach to First Philosophy." In *Perspectives on Embodiment: The Intersections of Nature and Culture,* edited by Gail Weiss and Honi Fern Haber, 121–41. New York: Routledge, 1999.

Sappol, Michael. *A Traffic in Dead Bodies: Anatomy and Embodied Social Identity in Nineteenth-Century America.* Princeton, NJ: Princeton University Press, 2002.

Sauer, Beverly. *The Rhetoric of Risk: Technical Documentation in Hazardous Environments.* Mahwah, NJ: Lawrence Erlbaum Associates, 2003.

Saunders, Barry F. *CT Suite: The Work of Diagnosis in the Age of Noninvasive Cutting.* Durham, NC: Duke University Press, 2008.

Schiappa, Edward. *The Beginnings of Rhetorical Theory in Classical Greece.* New Haven: Yale University Press, 1999.

Schmitt, Brandi. "A message from the Anatomical Services Committee," 31 May 2013, aaca@lists.aecom.yu.edu.

Schulte-Sasse, Linda. "Advise and Consent: On the Americanization of *Body Worlds.*" *BioSocieties* 1, no. 4 (2006): 369–84.

Schuster, Mary Lay. "A Different Place to Birth: A Material Rhetoric Analysis of Baby Haven, a Free Standing Birth Clinic." *Women's Studies in Communication* 29, no. 1 (2006): 1–38.

Schwenger, Peter. "Corpsing the Image." *Critical Inquiry* 26, no. 3 (Spring 2000): 395–413.

Scott, J. Blake. *Risky Rhetoric: AIDS and the Cultural Practices of HIV Testing.* Carbondale, IL: Southern Illinois University Press, 2003.

Segal, Judy Z. *Health and the Rhetoric of Medicine.* Carbondale, IL: Southern Illinois University Press, 2005.

Seidman, Irving. *Interviewing as Qualitative Research: A Guide for Researchers in Education and the Social Sciences.* 3rd ed. New York: Teachers College Press, 2006.

Sekula, Allan. "On the Invention of Photographic Meaning." In *Photography Against the Grain: Essays and Photo Works, 1973–1983,* edited by Allan Sekula, 1–21. Halifax, Canada: Press of the Nova Scotia College of Art and Design, 1984.

Sharp, Lesley A. *Strange Harvest: Organ Transplants, Denatured Bodies, and the Transformed Self.* Berkeley, CA: University of California Press, 2006.

Skloot, Rebecca. "Taking the Least of You." *New York Times Magazine,* April 16, 2006. www.nytimes.com/2006/04/16/magazine/16tissue.html?pagewanted=all&_r=0.

Slack, Jennifer Daryl, David James Miller, and Jeffrey Doak. "The Technical Communicator as Author: Meaning, Power, Authority." In *Power and Legitimacy in Technical Communication, Volume 1: The Historical and Contemporary Struggle for Professional Status*, edited by Teresa Kynell-Hunt and Gerald Savage, 169–92. Amityville, NY: Baywood Publishing, 2003.

Sontag, Susan. *On Photography*. New York: Picador, 1977.

Spinuzzi, Clay. *Tracing Genres through Organizations: A Sociocultural Approach to Information Design*. Cambridge, MA: MIT Press, 2003.

Spinuzzi, Clay. *Network: Theorizing Knowledge Work in Telecommunications*. New York: Cambridge University Press, 2008.

Stewart, John, Olivier Gapenne, and Ezequiel A. Di Paolo, eds. *Enaction: Toward New Paradigm in Cognitive Science*. Cambridge, MA: MIT Press, 2010.

Sullivan, Dale. "A Closer Look at Education as Epideictic Rhetoric." *Rhetoric Society Quarterly* 23, no. 3/4 (1993a): 70–89.

Sullivan, Dale. "The Ethos of Epideictic Encounter." *Rhetoric Review* 11, no. 2 (1993b): 113–33.

Sullivan, Dale L. "The Epideictic Rhetoric of Science." *Journal of Business and Technical Communication* 5, no. 3 (1991): 229–45.

Swarts, Jason. "Information Technologies as Discursive Agents: Methodological Implications for the Empirical Study of Knowledge Work." *Journal of Technical Writing and Communication* 38, no. 4 (2008): 301–29.

Tardy, Christine. *Building Genre Knowledge*. Anderson, SC: Parlor Press, 2009.

Tebeaux, Elizabeth. *The Emergence of a Tradition: Technical Writing in the English Renaissance, 1475–1640*. Amityville, NY: Baywood Publishing, 1997.

Thompson, Evan. *Mind in Life: Biology, Phenomenology, and the Sciences of Mind*. Cambridge, MA: Harvard University Press, 2010.

Timmerman, David M. "Epideictic Oratory." In *Encyclopedia of Rhetoric*, edited by Theresa Enos, 228–31. New York: Garland, 1996.

Timmerman, David M., and Edward Schiappa. *Classical Greek Rhetorical Theory and the Disciplining of Discourse*. New York: Cambridge University Press, 2010.

Titmuss, Richard. *The Gift Relationship: From Human Blood to Social Policy*. Edited by Ann Oakley and John Ashton. London: London School of Economics, 1997. First published 1970 by Allen and Unwin.

Tufte, Edward. *Visual Explanations: Images and Quantities, Evidence and Narrative*. Chesire, CT: Graphics Press, 1997.

Tufte, Edward. *The Visual Display of Quantitative Information*. Chesire, CT: Graphics Press, 2001.

Tufte, Edward. *Beautiful Evidence*. Chesire, CT: Graphics Press, 2006.

Turner, Mark. *The Literary Mind*. New York: Oxford University Press, 1996.

Uniform Law Commission. "Anatomical Gift Act (2006) Summary." The National Conference of Commissioners on Uniform State Laws. 2011. Accessed June 22, 2011. http://uniformlaws.org/Act.aspx?title=Anatomical+Gift+Act+%282006%29

Valverde de Amusco, Juan. *Historia de la composicion del cuerpo humano*. Rome: Ant. Salamana and Antonio Lafrery, 1560.

Van Dijck, Jose. *The Transparent Body: A Cultural Analysis of Medical Imaging*. Seattle: University of Washington Press, 2005.

Van Ittersum, Derek. "Distributing Memory: Rhetorical Work in Digital Environments." *Technical Communication Quarterly* 18, no. 3 (2009): 259–80.

Varela, Francisco J. "The Reenchantment of the Concrete." In *Incorporations*, edited by Jonathan Crary and Sanford Kwinter, 320–38. New York: Zone Books, 1992.

Varela, Francisco J., Evan Thompson, and Eleanor Rosch. *The Embodied Mind: Cognitive Science and Human Experience*. Cambridge, MA: MIT Press, 1991.

Vesalius, Andreas. *De humani corporis fabrica*. Basel: Joannes Oporinus, 1543.

Von Hagens, Gunther, Klaus Tiedemann, and Wilhelm Kriz. "The Current Potential of Plastination." *Anatomy and Embryology* 175, no. 4 (1987): 411–21.

Waldby, Catherine. *The Visible Human Project: Informatic Bodies and Posthuman Medicine*. New York: Routledge, 2000.

Waldby, Catherine. "Stem Cells, Tissue Cultures and the Production of Biovalue." *Health: An Interdisciplinary Journal for the Social Study of Health, Illness and Medicine* 6, no. 3 (2002): 305–23.

Waldby, Catherine and Robert Mitchell. *Tissue Economies: Blood, Organs, and Cell Lines in Late Capitalism*. Durham, NC: Duke University Press, 2006.

Waldby, Catherine, and Susan Merrill Squier. "Ontogeny, Ontology, and Phylogeny: Embryonic Life and Stem Cell Technologies. *Configurations* 11, no. 1 (2003): 27–46.

Walker, Jeffrey. *Rhetoric and Poetics in Antiquity*. New York: Oxford University Press, 2000.

Walker, Jeffrey. *The Genuine Teachers of This Art: Rhetorical Education in Antiquity*. Columbia, SC: University of South Carolina Press, 2011.

Walter, Tony. "*Body Worlds*: Clinical Detachment and Anatomical Awe." *Sociology of Health and Illness* 26, no. 4 (2004): 464–88.

Warner, John Harley, and Lawrence J. Rizzolo. "Anatomical Instruction and Training for Professionalism from the 19th to the 20th Centuries." *Clinical Anatomy* 19, no. 5 (2006): 403–14.

Warner, John Harley, and James M. Edmonson. *Dissection: Photographs of a Rite of Passage in American Medicine, 1880–1930*. New York: Blast Books, 2009.

Wartofsky, Marx W. "Picturing and Representing." In *Perception and Pictorial Representation*, edited by Calvin F. Nodine and Dennis F. Fisher, 272–83. New York: Praeger, 1979.

Washington, Harriet A. *Medical Apartheid: The Dark History of Medical Experimentation on Black Americans from Colonial Times to the Present*. New York: Harlem Moon, 2006.

Wells, Susan. *Out of the Dead House: Nineteenth-Century Women Physicians and the Writing of Medicine*. Madison, WI: University of Wisconsin Press, 2001.

Winsor, Dorothy. *Writing Power: Communication in an Engineering Center*. Albany: State University of New York Press, 2003.

Wollheim, Richard. *On Drawing an Object*. London: Lewis and Company, 1965.

Wysocki, Anne. "Impossibly Distinct: On Form/Content and Word/Image in Two Pieces of Computer-Based Multimedia." *Computers and Composition* 18, no. 2 (2001): 137–62.

Wysocki, Anne Frances. "The Multiple Media of Texts: How Onscreen and Paper Texts Incorporate Words, Images, and Other Media." In *What Writing Does and How It Does It: An Introduction to Analyzing Texts and Textual Practices*, edited by Charles Bazerman and Paul Prior, 123–63. Mahwah, NJ: Lawrence Erlbaum Associates, 2004.

Wysocki, Anne Frances, Johndan Johnson-Eilola, Cynthia L. Selfe, and Geoffrey Sirc. *Writing New Media: Theory and Applications for Expanding the Teaching of Composition*. Logan, UT: Utah State University Press, 2004.

Young, Katharine. *Presence in the Flesh: The Body in Medicine*. Cambridge, MA: Harvard University Press, 1997.

INDEX